California
Butterflies

California Natural History Guides: 51

California
Butterflies

John S. Garth
and
J. W. Tilden

Illustrated by
David Mooney and Gene M. Christman

UNIVERSITY OF CALIFORNIA PRESS
Berkeley Los Angeles London

CALIFORNIA NATURAL HISTORY GUIDES
Arthur C. Smith, *General Editor*

Advisory Editorial Committee:
Raymond F. Dasmann
Mary Lee Jefferds
A. Starker Leopold
C. Don MacNeill
Robert Ornduff
Robert C. Stebbins

University of California Press
Berkeley, California
University of California Press, Ltd.
London, England

Printed in the United States of America
10 9 8 7 6 5 4 3 2 1

This book is printed on acid-free paper to ensure
its permanence and durability.

Library of Congress Cataloging in Publication
Data
Garth, John S. (John Shrader), 1909–
    California butterflies.
    (California natural history guides ; 51)
    Bibliography: p.
    Includes index.
    1. Butterflies—California—
Identification.   2. Insects—Identi-
fication.   3. Insects—California—
Identification.   I. Tilden, J. W.
II. Title.   III. Series.
QL551.C3G37   1985   595.78'9'09794
        84-28071
ISBN 0-520-05249-8
ISBN 0-520-05389-3 (pbk.)

Dedicated to

the past, present, and future

Lepidopterists of our State

On Discovering a Butterfly

I found it in a legendary land
All rocks and lavender and tufted grass,
Where it was settled on some sodden sand
Hard by the torrents of a mountain pass.

I found it and I named it, being versed
In taxonomic Latin; thus became
Godfather to an insect and its first
Describer—and I want no other fame.

—Vladimir Nabokov, Quoted in
The Magnificent Foragers
(Smithsonian Institution)

CONTENTS

ACKNOWLEDGMENTS

The authors wish to thank the California Academy of Sciences, Paul H. Arnaud, Jr., Curator, for many courtesies extended; the Natural History Museum of Los Angeles County, the late Lloyd M. Martin and Julian P. Donahue, past and present Curators, for the loan of specimens not otherwise available, and for the privilege of copying original paintings of life history stages by the late C. M. Dammers. For critical review and many helpful suggestions the authors are indebted to John Lane, Santa Cruz City Museum; C. Don MacNeill, Oakland Museum; and Arthur C. Smith, General Editor of this series. Text illustrations in the introductory chapters are by Gene M. Christman; color and black-and-white plates are the work of David Mooney.

The authors also wish to thank the many colleagues, too numerous to mention individually, who have, over the years, contributed valuable information and suggestions.

INTRODUCTION

Of the 763 species of butterflies and skippers listed in the 1981 *Catalogue/Checklist of the Butterflies of America North of Mexico*, by Miller and Brown, at least 235 have been found in California, and another has been added since that date. All of the families of Nearctic butterflies and skippers occur here, although the Libytheidae and the Megathymidae are not represented in northern California.

This diversity may be accounted for by the great variety of habitats present. The state contains a coastal belt, a group of offshore islands, a central valley (Sacramento–San Joaquin), a high desert (Mojave), a low desert (Colorado), and several mountain ranges, of which the Sierra Nevada, with elevations of over 14,000 ft., is the most extensive. The Transverse and Peninsular ranges of southern California, with elevations of over 12,000 ft., have no truly alpine butterflies.

With the help of this field guide, users may learn where to find the butterflies and skippers of California, which range in size from the tiny Pygmy Blue, with a wing span sometimes as small as half an inch, to the Two-tailed Swallowtail, which may have a wing expanse of half a foot. Students may learn how to observe these Lepidoptera, how to study their fascinating habits, how to record observations so they will have meaning for others, and how to collect and preserve specimens indefinitely. Readers may also learn where to turn for additional information by consulting references such as those listed at the end of this book.

Information is given for all species known to have been recorded from California, regarding habitat, distribution, and abundance of adults; flight periods; and, if known, early stages and larval food plant preferences. Nearly all species and many subspecies are illustrated in color or, when color is not essential to identification, in black and white. In selecting subspecies for illustration, an effort has been made to achieve a

The principal physical features of California. By E. W. Jameson, Jr., adapted from E. W. Jameson, Jr., and Hans J. Peeters, *California Mammals* (Berkeley and Los Angeles: University of California Press, forthcoming).

balanced representation from all parts of the state. A checklist giving both scientific and common names is included.

With but very few exceptions, the names used in this book are those of L. D. Miller and F. M. Brown from the 1983 *Check List of the Lepidoptera of America North of Mexico*, edited by Ronald W. Hodges et al. and published in London by E. W. Classey, Ltd., and the Wedge Entomological Research Foundation.

The fifty-eight counties of California. By E. W. Jameson, Jr., adapted from E. W. Jameson, Jr., and Hans J. Peeters, *California Mammals* (Berkeley and Los Angeles: University of California Press, forthcoming).

Although the list of butterflies and skippers inhabiting California may be considered to be essentially complete, there are many gaps in our knowledge of the habits and life histories of individual species. The Pacific Coast Entomological Society, meeting at the California Academy of Sciences in San Francisco, provides for the exchange of such information on an informal basis. The Lorquin Entomological Society meets at the Natural History Museum of Los Angeles County. A group known as "Los Entomologos" (the Entomologists) meets at the San Diego Natural History Museum, and the Santa Barbara Natural History Museum sponsors a similar group. Thus even the beginner may make a contribution to our knowledge of this conspicuous and accessible group of insects.

1 · STRUCTURE, BEHAVIOR, AND DISTRIBUTION

Butterfly Structure and Growth

How a Butterfly's Body is Organized

The bodies of insects, including butterflies, are composed of ringlike sections called segments. The body of a butterfly, like that of other insects, is formed of three sections—head, thorax, and abdomen. The segments of the abdomen are easily visible. Those of the head and thorax are largely fused and are much less easy to distinguish. The appendages of the insect body, including mouthparts, legs, and wings, occur in pairs, one pair to each segment that bears them.

The head bears the mouthparts and the antennae (feelers), as well as the many-faceted compound eyes, which are not considered to be appendages. The mouthparts of a butterfly consist of a coiled tube, the proboscis, a paired structure in which the halves fuse, as shown in Figure 1. This structure, by which the butterfly sucks up liquids, lies between two scaly or hairy protruding structures, the palpi. The eyes occupy much of the sides of the head. The eyes of butterflies are good at perceiving movement and are also able to see color, including ultraviolet, which man cannot see.

The thorax, the section next behind the head, is formed of three segments, each of which bears a pair of legs. The thoracic legs are jointed, as are those of all arthropods, a group of organisms that includes spiders, crabs, scorpions, and others, as well as insects. The parts of a usual butterfly leg are shown

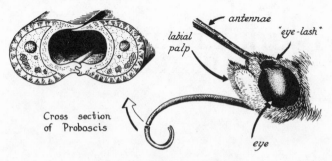

FIG. 1 Head with named parts

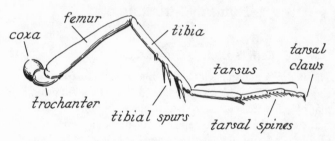

FIG. 2 Leg with named parts

and named in Figure 2. The second and third thoracic segments each bear a pair of wings as well. The front wings are usually much larger than the hind wings and have somewhat different venation.

The abdomen of the adult does not bear either legs or wings. At the tip of the abdomen are located the external reproductive organs, which are the claspers (valves) in the male, and a transverse opening in the female. The abdomen of the male is more slender than that of the female, especially if the female is gravid (full of eggs).

Inside the abdomen are located the digestive organs and the internal reproductive organs, the testes in the male, the ovaries in the female.

Along the sides of the body are segmental openings called

spiracles, which lead into a series of internal tubes called tracheae, which bring air directly to the internal organs.

The larva, or caterpillar, is very different in form from the adult. The mouthparts are chewing jaws (mandibles) and a "lower lip" (labium) containing a spinneret, from which is spun the silk used by the larva at pupation. The eyes consist of several simple eyes on each side of the head; there are no compound eyes. The three thoracic segments bear the true legs, which are quite short. The third, fourth, fifth, sixth, and last abdominal segments bear what are called prolegs. These function as legs, and have at their outer ends small hooks called crochets, which help the caterpillar in climbing. These are lost when the insect sheds its last larval skin and becomes a pupa, or chrysalis. The structures of a caterpillar are shown in Figure 3.

How a Butterfly Grows

While man and other vertebrate animals grow, develop, and reproduce without great change in bodily form, many invertebrates, including butterflies, devote an entire stage of their life to each of these activities, with successive changes in body form. The caterpillar, hatching from the egg, feeds and grows, storing up energy for the next stages. The quiescent pupa, or chrysalis, is a reorganizational stage, from which the adult emerges. The adult is the reproductive stage. It feeds on fluids only, the energy from which supports reproductive activities.

The eggs of butterflies are of many forms. Four are shown

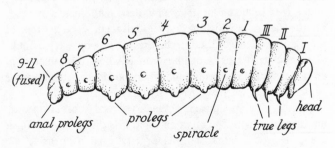

FIG. 3 Caterpillar with named parts

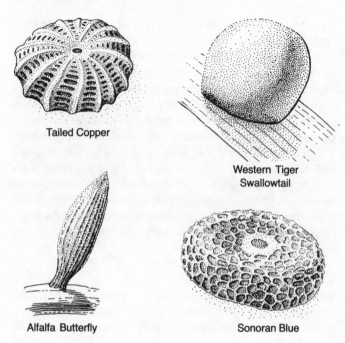

Tailed Copper

Western Tiger
Swallowtail

Alfalfa Butterfly

Sonoran Blue

FIG. 4 Types of eggs

in Figure 4. Most butterfly eggs have minute grooves, pits, or ridges on their surface, a condition spoken of as sculpture. The eggs are deposited by the female either singly or in clusters, usually on twigs, leaves, or buds of the plant on which the caterpillar feeds. In some cases, such as that of the fritillaries, which feed in the larval stage on violets, the female may never see the plants on which her larvae will feed, since the tops of the plants will have dried up before she emerges as an adult. She deposits her eggs on the ground where the violets will grow the following spring, and when the larvae hatch from the eggs, they find the new growth by themselves.

The chewing mouthparts of the larva work from side to side, not up and down. The larva eats and fills out its body until it seems it would burst. This is exactly what happens to the tightly packed outer skin, but a new and more flexible skin has been forming under the old one. When the old skin splits, the larva wriggles out of it and expands to a much larger size while the new skin is still able to stretch. This skin soon becomes hardened, and the larva resumes feeding until the next change of skin takes place. Each change of skin is called a molt. The time between each molt is called a stadium, and the larva itself at each stage is called an instar. Caterpillars pass through several instars before reaching the next change in form, pupation.

When the larva has reached the maximum size for its species, it crawls away, often entirely away from the food plant, and finds a sheltered place, where it will change into the pupa, in butterflies also called a chrysalis. Different families of butterflies pupate in different ways. The pupa described here and shown in Figure 5 is of the Monarch or Milkweed Butterfly.

On the lower lip, or labium, of the larva is a tiny tube, the spinneret, from which liquid silk, from a silk gland, can be spun. From the spinneret, the larva first weaves a button of silk. As the last larval skin splits, it frees the tip of the pupal abdomen. This tip has a cluster of small spines, the cremaster, and these spines hook into the silk button, so that the pupa now hangs head down, suspended by the cremaster only.

Prepupa

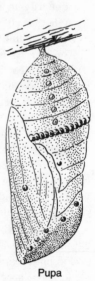

Pupa

FIG. 5 Pupation of Monarch

Swallowtails and members of the family Pieridae are also attached to the cremaster at the tip of the abdomen, but with the head up and leaning against a supporting silken belt. Some Satyridae and the parnassians pupate at the surface of the ground or in shallow earthen cells. Lycaenidae and Riodinidae usually pupate flat to a surface, but still have the anal cremaster and often a silken belt also. Skippers usually form a slight silken shelter around themselves, whereas giant skippers usually pupate in burrows in their food plants. The pupa resembles neither the caterpillar from which it developed nor the adult which will emerge from it. It is the reorganizational stage inside of which the larva changes into the adult.

After a period of as little as two weeks in some many-brooded species, or as long as a year or more in certain single-brooded species, the final change from pupa to adult, or imago, takes place. The chrysalis ruptures, and the winged insect slowly emerges. At first the wings are small; then as blood is pumped through their veins, they lengthen and hang loosely. Further inflation brings them to full size. The flow of blood stops, the wings become firm and stiffen until able to support the butterfly in flight. Meanwhile the butterfly has crawled from its place of concealment to an exposed place from which to leave for sunlit skies.

The process by which a caterpillar transforms into a butterfly is called metamorphosis, a word meaning "change in form." And since a pupa, or chrysalis ("resting stage") is involved, such metamorphosis is said to be complete, as distinguished from incomplete metamorphosis, exemplified by the grasshopper, in which the young resembles the adult except for functional wings.

Butterflies may overwinter in various stages of development. Some overwinter in the egg, hatching as first instar larvae the following spring. In some, the egg hatches during the summer, and the hatchling larva overwinters, becoming active at the end of winter. Others hatch from the egg shortly after the egg is laid, then feed for one or more instars, enter a resting period until the following spring, and then complete development. Some species, such as some of the whites and swallowtails, overwinter as pupae and emerge in the spring. Still others, as the anglewings and the Mourning Cloak, overwinter as adults and lay eggs the next spring. Some species, particularly desert species, may remain dormant in some stage of development for several years until sufficient rain falls to ensure their development through the growth of their food plants.

How to Rear a Butterfly

Butterflies that have flown for some time become worn. To obtain perfect specimens it is desirable to have butterflies that have never flown. This may be done by rearing them. You may follow a female as she flits from place to place, depositing an egg at a time on the chosen food plant. A gravid female may be confined with the food plant and induced to lay her eggs with offerings of flowers or sweetened water. Or you may search a known food plant for the larvae or eggs, or the immediate environs for the pupae, which need only to be kept until the adult emerges. If the larva is to be reared, an assured supply of the food plant is needed.

If a growing plant is used, the caterpillars may be enclosed in a gauze sleeve on the plant. If branches of the plant are used, a variety of cages may be made—shoe boxes or ice cream cartons in which cellophane windows have been made work well. Or a small plant or branch may be placed in a flower pot and covered with a lamp chimney. Examples of rearing cages are shown in Figure 6.

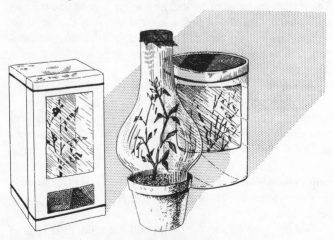

FIG. 6 Rearing cages

If grass or small clippings are to be used, they may be placed in a disposable paper or plastic cup covered with gauze secured by a rubber band. Since cleanliness is important, the cages should be cleaned often, and a fresh cup should be used with each food change. Larvae kept under soiled conditions may develop diseases and die.

Be careful to place your plant cuttings in moist sand, not water, because the caterpillars might crawl into the water and drown.

Small larvae may be transferred from old food to new using a small camel's hair brush.

Some butterflies do not pupate on the plant and need to be supplied with dead twigs or litter.

Records should be kept of the food plants used, the dates of pupation and emergence, and any unusual observations made.

If you wish to preserve one of the larvae, this may be done by killing it in hot water and preserving it in 70 percent alcohol. The accompanying label should be written in soft lead pencil and inserted into the vial with the specimen. The technique for inflating a caterpillar skin over hot air and mounting it dry on an insect pin is given by Holland (see References at end of book).

How Butterflies Behave

The total knowledge of butterfly behavior is great and is being added to continually. In such a book as this only the barest essentials can be mentioned. It is hoped that the interested student will be stimulated to examine the extensive literature on the subject.

As we have seen, butterflies have four stages of development. Each stage behaves differently from the others, and it has been said that each behaves as though it were an independent organism. The egg produces the larva. The larva feeds and grows to a specific size. The pupa, or chrysalis, reorganizes the larval form into the form of the adult. The adult mates and reproduces, laying eggs at a time and place suitable for the perpetuation of the species. Each stage has its appropriate action.

The egg (see Figure 4). Some butterflies lay their eggs in masses; others deposit a few eggs in one place. Some eggs are laid one to a place on the same plant. Others are laid one to a plant, after which the female flies to another plant before depositing the next egg. In some species the larva develops in the egg in a short time and hatches. In others, the development of the larva is delayed for some time. In yet others, the larva develops in the egg quite soon, but remains inside the egg shell until the conditions are right for it to hatch. A few examples follow.

The Bay Region Checkerspot lays its eggs in masses, on or near the food plants. Fritillaries of the genus *Speyeria* lay their eggs in the vicinity of the dry food plants, or even drop their eggs from the air.

The Fiery Skipper often lays its eggs on lawn grasses, carefully placing each egg on the underside of a grass leaf. Blues of the genus *Euphilotes* lay their eggs on the same flower heads from which the adults have nectared, and on which the larvae develop. They are truly one-plant insects, the plants being various species of wild buckwheat (*Eriogonum*).

Perhaps the most remarkable, some populations of Lindsey's Skipper lay their eggs on a lichen (*Usnea florida*) that grows on tree trunks and wooden fences. The larva emerges in the spring, makes its way down to the ground, and finds its own food plants, which are Blue Bunch Grass (*Festuca idahoensis*) and California Oat Grass (*Danthonia californica*).

The larva. Larvae (Figure 7) may be brightly colored or inconspicuous. Brightly colored larvae are often distasteful to would-be predators and are said to show warning coloration. Inconspicuous (cryptic) larvae are usually edible, and their dull coloration provides some measure of protection. Some larvae live in shelters of their own making; others hide in duff on the ground. Some feed openly in the daytime; others hide by day and feed by night.

Larvae may feed on a single species of food plant, or on several related plants, or on unrelated plants that have a similar chemical makeup, or, as do some species, on a wide variety of unrelated food plants.

The larvae of some species of butterflies develop directly, feeding and growing continuously until full sized. Other spe-

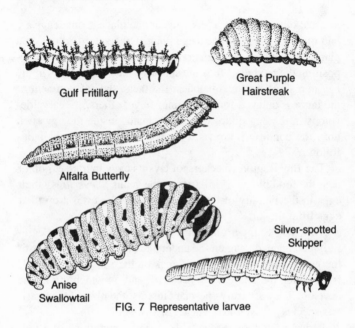

Gulf Fritillary

Great Purple
Hairstreak

Alfalfa Butterfly

Anise
Swallowtail

Silver-spotted
Skipper

FIG. 7 Representative larvae

cies feed and grow for a while, then enter a state of suspended activity (diapause) that will carry them over an unfavorable part of the year. Usually the diapause is in winter, but some species have a summer diapause, their food being too dry or tough at that time. Many species overwinter as hatchlings and develop the following spring.

The Monarch caterpillar is ringed with yellowish and greenish gray, and is distasteful. It feeds openly on its food plants, various milkweeds (*Asclepias*), and often pupates on the same plant. These caterpillars usually feed at some distance from one another.

The West Coast Lady caterpillar feeds on many members of the Mallow Family (Malvaceae). It makes a shelter by drawing the edges of the leaf together with silken threads. It does not have definite broods. The larvae can be found on the plants during most of the year. Larvae may be present in moderate climates even in winter and will feed whenever the weather is warm.

The blackish caterpillar of the Common or Chalcedon

Checkerspot feeds openly on various members of the Figwort Family (Scrophulariaceae). After hatching in the late spring or early summer, the young larvae feed for a while until their food plants begin to dry up, then enter a long overwintering diapause and resume feeding the following spring.

The caterpillar of the Woodland Skipper hatches from the egg in the fall and overwinters as a hatchling. The caterpillar lives in a silken tube in a rolled-up leaf of tall woodland grasses. It reaches full size in late May, goes into a summer diapause until late July or early August, then pupates and soon emerges as an adult.

The pupa (chrysalis) (Figure 8) is usually cryptically colored, and is often difficult to see. The method of attachment of the members of the various families was described above.

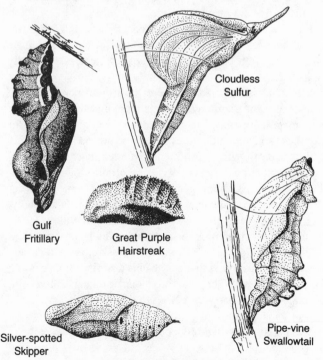

Cloudless Sulfur

Gulf Fritillary

Great Purple Hairstreak

Pipe-vine Swallowtail

Silver-spotted Skipper

FIG. 8 Representative pupae

The chrysalis of the Anise Swallowtail may be either brown or green. The adult may emerge the same season, or may wait one or two years before emerging. The chrysalis of the Common Checkerspot is light gray marked with many tiny dark lines and dots. It hangs either on its food plant or on nearby vegetation. The chrysalis of the Gulf Fritillary is strange looking, the abdomen slender, the anterior part wide and flattened. It can twist slightly, possibly to present its narrower edge to direct sunlight.

The adult (imago). Adult butterflies recognize one another during courtship and mating by color, some combination of markings, manner of movement, and odor. Different species emit different attractant odors (pheromones). The mating habits of various species differ widely. Some butterflies mate early in the day, other species at times ranging from midmorning to early evening. Of the many strategies by which butterflies meet and mate, five are mentioned here.

Bay Region Checkerspot males course back and forth over a field where females will emerge and discover and mate with them, often before the wings of the females have fully expanded.

Males of Lorquin's Admiral perch on a stick or branch and at intervals fly out, returning to the same or a nearby perch. The same male will occupy the same territory for days, perhaps for life. Any large insect that flies by will be pursued. Almost any kind of butterfly will be followed, and often touched, but mating is attempted only when a female Lorquin's Admiral is encountered. A male may pursue a female for some distance. Odor is most likely the final factor in recognition. Incidentally, it is amusing to note that male admirals will follow to the ground a chip, small rock, or clod thrown by an observer.

The male Buckeye is even more aggressive. He will closely pursue any large insect that crosses his territory, often chasing it out of the area. When a receptive female enters his territory, the male follows her very closely. Often the pair may flutter high into the air, returning to the ground to mate.

The female Woodland Skipper sits quite openly, usually on the ground, and vibrates her wings at a fixed rate. A passing male will make several flights over her, approaching closely.

Eventually she tips her abdomen up, and it is possible that a pheromone is released at this time. The mating that ensues lasts for some time. If disturbed, the pair flies away together, still attached. Other males may be attracted; one sometimes sees three or four males flying back and forth over a mating pair.

The Propertius Dusky-wing is our commonest and largest dusky-wing. Its larva feeds on oaks, making a shelter in the leaves. Adults are commonly seen in openings in oak woodland. The males course a few feet above the ground, searching for females. When a pair meets, an extended nuptial flight may occur, the skippers often fluttering to a considerable height.

After mating, a female butterfly will search for a proper plant on which to lay her eggs. The odor of the plant seems to be a major factor in its selection. However, instances are known in which a plant is attractive to a female butterfly, but the plant is a poor or unsuitable host for the larvae.

Adult butterflies take only liquid food—usually nectar obtained from flowers. Some butterflies visit many kinds of flowers. Others are more specific in their choices. The Common Checkerspot will take nectar at almost any kind of flower. Blues of the genus *Euphilotes* seldom visit any kind of flower except those of the species of *Eriogonum* on which their larvae feed.

Some butterflies get liquids from sources other than flowers. Some feed at sap flows, where the bark of a tree has been broken and sap exudes. Our Red Admiral often does this. Some butterflies eat the honeydew produced by aphids, a very nutritious food. Butterflies of certain species will visit the feces of animals, or even the decaying bodies of dead animals. The California Dog-face has been found in stable yards feeding on horse manure. It has been suggested that such sources may supply salts and minerals, as well as moisture.

Butterflies often visit mud puddles and sandbars, where they can be seen taking moisture. Sometimes a sandbar may be frequented by a number of swallowtails, all sitting with their heads facing the same direction. In hot weather, blues and hairstreaks have been observed to suck moisture, then expel a stream of water through the anus. This habit, which has been termed pumping, may be a way of cooling the inside of the body.

Many blues and hairstreaks sit with their wings brought together over their backs, and then rub the tips of their hind wings together. Many of these species have a brightly colored spot (the thecla spot), and sometimes other markings, at the tip (tornus) of each hind wing. It has been suggested that these markings, together with the rubbing movement, look like a moving head, and that predators may strike at the back end of the wings, giving the butterfly a chance to escape. This seems plausible, since such butterflies may now and then be found with the tips of the hind wings missing.

The caterpillars of blues and hairstreaks have glands at the anal end of the body that produce a secretion that is very attractive to ants. Ants often tend such larvae and will touch them at the end of the body, whereupon the larva extrudes a drop of fluid that is eaten by the ant. It is thought that the tending of the larvae by the ants may be a protection from predation, if not by other insects, at least from the ants.

Some butterflies make use of certain passageways between hills or high vegetation, and at times many individuals may use these flyways, as they are called. Collectors and observers often make use of these flyways to obtain specimens, or to note the species and number of individuals that pass by.

The coming together of individuals in one place is termed aggregation. The aggregation of butterflies under certain circumstances leads to types of group behavior of special interest. Two of these are hilltopping and migration.

Many species of butterflies have been found to play around the summits of hills or mountains. This behavior has been noted by many observers, and various explanations have been given. Oakley Shields, who made a detailed study of the phenomenon, showed beyond reasonable doubt that hilltopping is a way for butterflies to meet and mate. Males go to the tops of hills in considerable numbers, whereas females are less evident there. Captures show that unmated females come to the hilltops, mate, and return to lower elevations. Many species of butterflies have been found to hilltop, but swallowtails and Western Whites are especially conspicuous. On Mt. Hamilton, near San Jose, the Anise Swallowtail and the Western Tiger Swallowtail have been noted hilltopping by Tilden, as has also the Columbian Skipper, which has been observed several times in nuptial flights and mating.

Some species of butterflies may from time to time occur in great numbers, moving steadily in one direction. These movements are called migrations. The Monarch is the best known migratory butterfly. In very early spring, overwintering adults mate and then move gradually inland and northward, stopping along the way to oviposit on milkweed at times. After several generations, some adults reach inner continental areas and southern Canada, where another brood may be raised. In the fall, as the days grow shorter, the now sexually inactive adults, several generations removed from those that pushed east or north in the spring, gradually fly south and west, eventually to spend the winter in areas with climates mild enough for their survival. When spring comes, the adults will mate and the cycle will be repeated.

Through popular magazines such as *National Geographic*, almost everyone knows of the migrations of the Monarch. As mentioned, it may summer as far north as the northern United States and southern Canada. The western populations winter along coastal California; the eastern populations winter primarily in the highlands of Mexico.

Birds, of course, are much more extensively migratory than butterflies. Butterflies differ from birds in ways that affect the extent of their migrations. Butterflies are cold blooded, whereas birds are warm blooded. Butterflies are limited in activity to warm conditions, whereas birds are active in cold weather as well. Butterflies have a short life span, usually less than a year, often only a short season as an adult, so that the same individual cannot make both the northward and southward migration. Butterflies also have, as a rule, less extended powers of flight, so usually travel fewer miles in one day. However, some marked Monarchs have been found to have made total flights of over 2,000 miles (see Figure 9).

In California, Monarchs overwinter in groves of native Monterey Pine and in eucalyptus groves, from at least Marin County to San Diego County. In Pacific Grove there is an annual Butterfly Festival, and Monarch butterflies are protected by local ordinance. The University Extension at the University of California, Santa Cruz, and the Ventura Chamber of Commerce conduct organized tours each November and December to places where Monarchs congregate. It is quite an experience

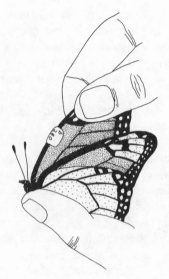

FIG. 9 Tagged Monarch

to step into a clearing in such a grove and see butterflies covering the branches more densely than leaves (see Figure 10). All are still except those in the topmost branches that have been touched by the first rays of sunlight, awakening them to daily flight. As the sun penetrates the clearing, more and more awaken to its warming touch. Soon the clearing is alive with darting forms, some alighting on the observers. Then, as the last rays of sunlight leave the treetops in late afternoon, all settle down again for the long winter night. They will remain in the area until spring comes in late February or early March.

It is doubtful that any of our other California butterflies have return movements, as the Monarch does, but several other species may engage in mass movements when their populations are high.

Outbreaks of the Painted Lady occur at irregular intervals, some years apart. Adults originating in southern, often desert, areas, move northward as far as southern Oregon or beyond. In such outbreaks there is no return movement. The adults continue northward until they die. Motorists crossing the Mojave or Colorado deserts report encountering them, flying north-ward by hundreds or thousands. The flight may continue for several days. No one knows for sure how far south such flights may start. Garth encountered them halfway down Baja Califor-nia several days before the flight reached Alta California.

Mass movements of the California Tortoise Shell take place in early summer in the Cascades, the Sierra Nevada, the Santa Cruz Mountains, and the San Bernardino Mountains of south-ern California. Adults appear in great numbers, flying from dawn to dusk, from eye level to treetop height, and attract a good deal of interest. Sometimes highways are covered with the bodies of those that have been struck by cars. Prior to a flight year, the gregarious larvae spin webs on Buck Brush or Deer Brush (*Ceanothus* spp.). In some years a second flight has been noted at higher elevations in the Sierra Nevada, and in the vicinity of Mt. Shasta.

When an opportunity to observe migrations or mass move-ments of butterflies occurs, notes may be made, with answers to such questions as: Where were the butterflies seen? Were they all of one species, or of several species? What were the first and last days of observation? How early and late in the day were they flying? In what direction were they heading? Were they flying with the wind, against the wind, or cross-wind? What was their maximum, minimum, and average height above the ground? What was the average speed in estimated feet per second? How many individuals passed a given point per min-ute? Such information is of interest not only to the observer but to others who study the subject.

Most moths fly at night, and nearly all butterflies fly in the daytime, but now and then individual butterflies are attracted to lights at night. The species flying to artificial lights have cer-tain things in common: they are usually common species; they are usually found in the daytime in the same places; only a few individuals of the species will be found at the lights.

Two families, the skippers (Hesperiidae) and the brush-footed butterflies (Nymphalidae), make up the largest number of those species taken at lights in California. Of skippers, the Farmer, the Woodland Skipper, the Eufala Skipper, the Umber Skipper, the Common Sooty-wing, and the Silver-spotted Skip-

FIG. 10 Overwintering Monarchs

per have been taken at light once to several times each. Of the Nymphalidae, the Red Admiral, the Painted Lady, the West Coast Lady, the Buckeye, and the Mourning Cloak have come to light. Other butterfly species observed at lights are the Monarch, the Alfalfa Butterfly, and the Echo Blue.

The previous pages give only a selected few examples of the many fascinating types of behavior known to occur in butterflies. It is hoped that the reader will be interested in pursuing this aspect of the subject further.

How Butterflies Depend on Plants

Butterflies depend on plants in many ways. A student of butterflies becomes to some extent a botanist as well as an entomologist.

When one speaks of the food plant of a butterfly, one speaks of the larval food plant, the plant on which the caterpillar feeds. This may be one particular species of plant, or a number of different species of plants of the same, or different, families. If a larva feeds on only one food plant, it is said to be monophagous ("one-feeder"). If it feeds on several species of plants, these usually being in the same family, it is said to be oligophagous (*olig* = "few"). If it feeds on many species of plants, often of many different families, it is said to be polyphagous (*poly* = "many").

Relatively few butterflies are monophagous in the larval

stage. One that is, as far as is known, is the Hermes Copper (*Hermelycaena hermes*). It feeds only on Redberry (*Rhamnus crocea*), a member of the Buckthorn Family (Rhamnaceae).

Most butterflies may be classed as oligophagous, feeding on plants of one family or genus. Thus the butterflies known as whites (genera *Pontia* and *Artogeia*, formerly *Pieris*), choose plants of the Mustard Family (Brassicaceae), a family that, in addition to many native species of mustards and similar plants, also includes a number of cultivated vegetables such as cabbage, cauliflower, broccoli, and brussels sprouts. Some, but not all, of the sulfurs (genus *Colias*) feed in the larval stage on members of the Pea Family (Fabaceae), which includes many native species but also cultivated peas, beans, alfalfa, and others. Some species of swallowtails (family Papilionidae) use members of the Parsley Family (Apiaceae) in the larval stage. This plant family includes a large number of native and weedy species, and also many vegetable and spice plants, such as parsley, carrots, celery, dill, parsnips, caraway, and others.

An advantage of knowing the plant families is that developing larvae may in many cases be switched from the wild plant on which they were found, and which may not be easily available, to a related garden or ornamental plant. For instance, the larvae of the Juniper Hairstreak will do equally well on garden cypress, since juniper and cypress are both members of the plant family Cupressaceae.

Butterflies also feed as adults, but only on liquid foods. The plants from which they take nectar may have no relationship to the larval food plants. Common garden plants that are attractive to butterflies include lantana, heliotrope, verbena, and various members of the Mint Family (Lamiaceae). Native plants that are especially attractive include yerba santa (*Eriodictyon* spp.), wall flower (*Erysimum* spp.), mock orange (*Philadelphus* spp.), Coyote Mint (*Monardella villosa*), California Buckeye (*Aesculus californica*), wild buckwheats (*Eriogonum* spp.), and many members of the Sunflower and Dandelion Family (Asteraceae).

Whereas most butterflies prefer to fly in sunlight, others fly or rest in the shade, at least at times. In the pinyon-juniper woodland of the drier parts of the West, the brownish or blackish satyrs (genus *Cercyonis*) dart in and out of shadows, virtually disappearing each time they flit into the shade. The Canyon Oak Hairstreak (*Habrodais grunus*), sometimes very common, is most active in the morning and late afternoon. During the hottest part of the day, large numbers of the butterfly may be found hiding in shady vegetation, presumably to avoid the heat.

Others, such as the anglewings (genus *Polygonia*), and certain other members of the family Nymphalidae, obtain concealment by perching on the bark of trees, which their undersides so closely resemble. Their bright upper sides are shown as they dart out from their sheltered perches.

What do butterflies do when it rains? Again, it is a tree, a shrub, or tall grass, that provides shelter from inclement weather.

Where do butterflies pupate? Except for those whose larvae drop off trees and shrubs to pupate among the litter on the ground, the larvae attach themselves to a twig, a branch, a tree trunk, or dead bark, before molting to the chrysalis. Usually they secure themselves by the anal end to a silken button, and in some families, by a silken girdle as well.

Butterflies are closely associated with plants in many other ways. Many desert plants have spines or thorns that may deter certain would-be predators. The yucca and agave plants that are the hosts of the giant skippers are of this type. Others, such as the nettles on which larvae of anglewings and Red Admirals feed, have stinging (urticating) hairs. Many trees are hosts to parasitic mistletoes, which are the larval food plants of certain hairstreaks. The list is endless.

It seems apparent that to understand the ways of butterflies, it helps to understand something of plants as well.

Butterfly Distribution

The butterflies found in California are derived from at least three faunal elements: northern or arctic species; tropical species; and species belonging more strictly to the western United States and associated with food plants largely from the Madro-Tertiary Flora—plants that developed in the Sierra Madre area of Mexico and spread widely in the western United States. For an account of the geological history of California flora, see Axelrod in Munz, *A California Flora*, pp. 5–8.

Holarctic and Northern Butterflies

One who collects butterflies in the Alps or the Pyrenees of Europe cannot fail to notice certain similarities to the butterflies of the Sierra Nevada and of the Rocky Mountains. There are parnassians, fritillaries, coppers, ringlets, and blues, among others, some extremely similar to their New World counterparts. This resemblance occurs because North America and Eurasia were most recently connected, not across the North Atlantic, but across the North Pacific by the Bering Land Bridge and the Aleutian Arc. Butterflies were able to cross at sea level from Siberia to Alaska and proceed down the Rocky Mountain and Sierra-Cascade chains, seeking higher elevations as they moved south. At about the level of the Kern River and Walker Pass, which mark the southern limit of the Sierra Nevada, the arctic butterflies could proceed no further. Thus there are no truly boreal, or far northern, butterflies in southern California.

It has taken North American lepidopterists some time to realize how closely related European and North American butterflies are. Whereas each continent has many species and even genera that are different, they also have many genera, and some species, in common. In one such instance, it was shown by Demorest Davenport that North American ringlets (genus *Coenonympha*) are (except for *C. haydenii*) all very closely related to the European *C. tullia*. He placed all of our ringlets except *haydenii* as subspecies of *C. tullia*. Recently each has been considered a separate species, but members of the *C. tullia* species group.

Several Californian butterflies are the same subspecies as, or a different subspecies of, European butterflies. Among these are the Red Admiral, the Painted Lady, the American Copper, Behr's and Sternitzky's Parnassians, and our three subspecies of the Northern Blue. Among genera shared by Europe, Asia, and North America are *Parnassius, Papilio, Colias, Oeneis, Coenonympha, Nymphalis, Polygonia, Ochlodes, Erynnis*, and others. Thus butterflies are one more group that demonstrates the unity of the northern faunas of Europe, Asia, and North America, known to biogeographers as the Holarctic Region.

Sonoran and Neotropical Butterflies

A large part of California's butterfly fauna may be spoken of as Sonoran. These butterflies are the same as, or closely related to, those of northwestern Mexico, and have as food plants members of the Madro-Tertiary Flora, as noted above.

The word *tropical* suggests rain forest, but the New World Tropics, and especially the Sonoran Region, have vast areas of thorn forest, open woodland, grassland, and desert. This flora has its attendant fauna, including butterflies.

This southern element is dominant in the desert regions of southern and eastern California, and follows the Colorado River through the Grand Canyon into southern Colorado. The entire family of giant skippers (Megathymidae) is restricted to the Mojave and Colorado deserts, in California. Of the family of metalmarks (Riodinidae) only the Mormon Metalmark (*Apodemia mormo*) extends into northern California. Skippers (Hesperiidae) are better represented in the southern part of the United States, the number of species decreasing with increasing latitude until in our northern tier of states only a few species remain.

Certain butterflies have followed the introduction of semi-tropical and tropical plants into California. The Gulf Fritillary (*Agraulis vanillae incarnata*), our only representative of the family Heliconiidae, has followed plantings of the passion vine from southern California to as far north as Redding in the Sacramento Valley. The Giant Swallowtail (*Papilio cresphontes*) has followed citrus trees from the Imperial Valley into central California at least as far north as Porterville in the San Joaquin Valley. The giant sulfurs, *Phoebis sennae marcellina*, the Cloudless Sulfur, and *P. agarithe*, the Large Orange Sulfur, have followed plantings of *Cassia* shrubs in southern California gardens and, at times, at least as far north as San Jose, Santa Clara County.

Island Butterflies

Separated from the mainland by some distance, the southern California Channel Islands are sufficiently remote to have a butterfly fauna that differs in part from the mainland, but not as much as do those of oceanic islands separated from a continent by hundreds or thousands of miles.

The first thing to notice is that many of the mainland species are missing from the islands. Their absence may be either because they never reached the islands, or because the islands do not provide the variety of habitat or the particular larval food plants that they require. The second thing to notice is that, in comparing island specimens to mainland specimens of the same species, the individuals of the island populations tend to be smaller and darker than mainland specimens of the same species. These differences are not true of large and powerful species, such as the Monarch and Painted Lady, which pass regularly from island to mainland, but of small, weak fliers that do not make the crossing often, or perhaps made it only once.

It is also possible, however, that the island butterfly species have existed since the islands were part of the mainland, and inbreeding has produced differences of size, color, and pattern between the now-isolated island populations and their mainland counterparts. This theory seems to hold true for the Island Checkerspot (*Occidryas editha insularis*), a population now restricted to Santa Rosa Island but formerly occupying a continuous range with the mainland populations of *O. editha* nr. *bayensis* of the Santa Monica Mountains, with which the northern Channel Islands were once continuous. Fossil bones of the Pygmy Elephant unearthed on Santa Cruz Island testify to this former land connection.

The southern Channel Islands, and particularly Santa Catalina Island, have a different geological history. Their former connection with the mainland is not so readily apparent. One endemic (native) species of Santa Catalina Island is the Avalon Hairstreak (*Strymon avalona*), a full species, not a subspecies, and found nowhere else. Other populations that have existed on the California Channel Islands long enough to become subspecifically distinct are the orange-tips *Anthocharis cethura catalina*, the Catalina Orange-tip, found only on Santa Catalina Island, and *A. sara gunderi*, Gunder's Orange-tip, found on both Santa Catalina and Santa Cruz islands. One can only speculate on what may have been the endemic butterfly population of San Clemente Island, southernmost of the group, as the native vegetation has been decimated by goats introduced during the past century, and with it the butterflies that either fed upon the plants as larvae or nectared at them as adults.

Disappearing Butterflies

Something sinister is happening to California's butterflies: If one returns to a locality in which a butterfly was found fifty, twenty-five, or even ten years ago, it may not be found there now. If the area has remained unaltered by man, the chances of still finding the butterfly are most likely very good. But if man has moved in and built houses, factories, highways, or farms, we will have to look elsewhere for our butterfly, if indeed it can be found at all.

The most severe fate that awaits a butterfly under these circumstances—outright extinction—has already occurred in several cases. Formerly found in abundance on the sand dunes of San Francisco, the Xerces Blue (*Glaucopsyche xerces*) was last seen in 1941 before becoming a victim of urbanization. The same fate had earlier, in the 1880s, overtaken the Sthenele Satyr (*Cercyonis sthenele sthenele*) and, apparently sometime in the 1930s, the Pheres Blue (*Icaricia icarioides pheres*). Both these species were found in the hills of San Francisco, including the Presidio. Other butterflies considered extinct are Strohbeen's Parnassian (*Parnassius clodius strohbeeni*), last recorded in 1958, and the Atossa Fritillary (*Speyeria adiaste atossa*) of the Tehachapi Mountains, apparently last seen sometime in the 1950s.

There is probably no lepidopterist who does not know of a local colony of a favorite butterfly that has ceased to exist because a subdivision, dam, or highway disturbed or obliterated its natural habitat. Where once a flourishing colony of the Coastal Arrowhead Blue (*Glaucopsyche piasus catalina*) was found at the intersection of Verdugo Road and Glendale Boulevard in Glendale, only supermarkets, gas stations, and parking lots exist today. In fact, the lupine fields have all but disappeared from the San Fernando Valley, and the Arrowhead Blue is restricted to a few acres in a canyon above it. A gigantic earth-fill dam obliterated the largest known southern colony of the Sonoran Blue (*Philotes sonorensis*) in San Gabriel Canyon. Fortunately smaller colonies in adjacent canyons remain.

Because of the recognized threat to their continued existence, a number of California butterflies have been placed on the endangered species list, giving them federal protection under the Fish and Wildlife Service. Species and subspecies so designated may not be collected. Unfortunately, since many of the places where they are found are on private land, there is no guarantee that their habitats will be protected.

The butterflies currently (1983) stated to be endangered are Lange's Metalmark (*Apodemia mormo langei*) of the Antioch sand dunes; the Lotis Blue (*Lycaeides idas lotis*) of the coastal peat bogs of Mendocino County; the San Bruno Elfin (*Incisalia mossii bayensis*), found only in the San Bruno Mountains of the San Francisco Peninsula; the Mission Blue (*Icaricia icarioides missionensis*), first described from Twin Peaks, but also found in the San Bruno Mountains; Smith's Blue (*Euphilotes enoptes smithi*) of coastal Monterey County; the El Segundo Blue (*Euphilotes battoides allyni*), now restricted to a couple of small spots in the El Segundo sand dunes; and the Palos Verdes Blue (*Glaucopsyche lygdamus palosverdesensis*), apparently restricted to the Palos Verdes Peninsula, Los Angeles County. Other species and subspecies under consideration as of this writing (1984) are the Callippe Fritillary (*Speyeria callippe callippe*), now found only in the San Bruno Mountains; the Bay Region Checkerspot (*Occidryas editha bayensis*) and the Bay Region Blue (*Euphilotes enoptes bayensis*), both found in the San Francisco Bay area. There are no doubt others as well.

In most cases it is habitat destruction rather than overcollecting that has been responsible for the insect's decline or extinction. It has been thought that perhaps the Strohbeen's Parnassian and the Atossa Fritillary may have been on their way out because of natural changes in their environment. However, in the case of Strohbeen's Parnassian, the last thriving colony was nearly wiped out by extensive summer home development. The Atossa Fritillary may have suffered from overgrazing, or from repeated years of drought.

The conscientious lepidopterist, however, will obey the law (P.L. 93–205, Endangered Species Act) and will refrain from collecting endangered or threatened species, and will restrict all collecting to the number of specimens needed except in the case of very common species. Rare butterflies are best observed in an undisturbed habitat. We hope that such moderate or reduced collecting will enable our butterflies to delight future generations.

2 · OBSERVING AND COLLECTING BUTTERFLIES

How to Make a Butterfly Collection

Much can be learned about butterflies by watching them, but, for detailed study, the serious student may sooner or later need to have actual specimens. Such specimens should be carefully preserved, and eventually donated to some university or museum at such time as the student may no longer wish to maintain a collection.

Butterflies may be collected with a minimum of equipment, which can either be made or purchased from a dealer. A net is the first requirement. It consists of a metal ring, a cloth bag, and a wooden handle. A stiff wire hoop, bent as shown in Figure 11, a nylon netting bag, and a 2– to 4–foot length of bamboo or lightweight dowel serve well. The ring should be at least 15 inches in diameter. The net bag should be twice that in depth and should be reinforced with heavy cloth where it touches the ring. If preferred, nets may be dyed green, which makes them less conspicuous, and less alarming to butterflies.

Several kinds of commercial nets are available from supply houses, including nets with longer handles, collapsible nets that will fit into a suitcase, and even pocket nets of spring steel that when folded are only a few inches in diameter but become useful nets when opened.

The standard killing jar (Figure 12) is the cyanide bottle. However, cyanide is a deadly poison and should not be used by children nor kept where children or unsuspecting adults can

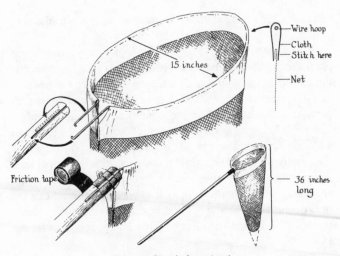

15 inches

Wire hoop
Cloth
Stitch here

Net

Friction tape

36 inches long

FIG. 11 Simple insect net

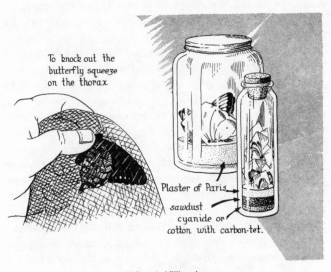

To knock out the
butterfly squeeze
on the thorax

Plaster of Paris

sawdust
cyanide or
cotton with carbon-tet.

FIG. 12 Killing jar

reach the collecting bottles. Cyanide is at times difficult to obtain.

As a collecting bottle, a pint or half-pint mayonnaise jar or similar container with a quarter-turn lid is suitable. Into this are placed a teaspoonful of cyanide crystals, a layer of sawdust carefully tamped down and a covering of liquid plaster of paris, separately mixed to the consistency of thin porridge. The jar is left uncovered until the plaster of paris has hardened and dried, perhaps as long as overnight or longer. The jar is then ready for use. For safety, it is a good idea to wrap at least the lower part of the jar with masking or plastic tape, to prevent shattering of the bottle should it chance to become cracked or broken. It is also a very good practice to label each cyanide jar with warnings—skull and crossbones and a warning, POISON.

Some other killing agents that may be used are ethyl acetate, ether, chloroform, and carbon tetrachloride, but it should be noted that all are poisonous or harmful to humans and should be treated carefully. If these liquid agents are used, the jar will need to be recharged before each use. Cyanide jars last for an entire season, or even longer.

A variant of the cyanide bottle is the cyanide can, made in the same way but using a hinge-top Band-Aid tin instead of a glass jar. A strip of paper cut the width of the can is folded like a fan and inserted, providing an individual pocket for each specimen. The can will not break, fits easily into the pocket, or may be attached to the belt. Since the can is not airtight, it should be taped shut when not in use (see Figure 13).

All of the collecting containers described above should be prepared where there is adequate ventilation, preferably outside, but at least on an open porch, or if in a laboratory, under an exhaust hood.

Another item needed for making a butterfly collection is a pair of tweezers or forceps (see Figures 15 and 18). These may be either straight or curved but should have an easy action because of the delicacy of a butterfly's wings. Some collectors prefer to carry a container with glassine envelopes, such as stamp collectors use, for enclosing the butterflies after capture. Such envelopes may be placed in a cyanide jar until the butter-

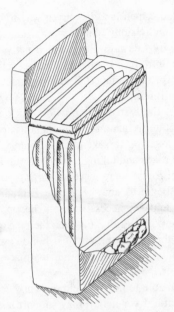

FIG. 13 Cyanide can

flies within them are dead, and then transferred to another container. If fresh butterflies are placed in glassine envelopes for indefinite storage, they may become moldy. It is therefore better to mount them at once, or allow them to dry a bit before storing them.

For field work, a shoulder bag is convenient for carrying the various items needed for collecting and taking proper care of specimens. If you are traveling by car, a box to hold all collecting equipment is most convenient.

Where to Find Butterflies

When California was largely open country, all that was necessary to find butterflies in abundance was to go into the nearby countryside. Every gulch or stream bed had a willow thicket in which Lorquin's Admiral and the Western Tiger Swallowtail could be found. Now that most urban and suburban and some rural streams are largely concretelined, many willows have been eliminated. Open land has given way·to clean-cultivated farms, or to subdivisions. Cities merge into one another and are connected by freeways along which native vegetation has given way to introduced ornamentals. Insecticides and herbicides have further reduced insect populations along such roadsides as still remain.

Fortunately, these freeways provide quick access to some more distant butterfly haunts, the mountains and deserts. Here the freeways must be forsaken for secondary highways, and these for back roads and hiking trails. The public campgrounds of our national forests may serve as starting points.

Butterflies begin to emerge in late February in our lower deserts, then appear successively on the upper deserts, the chaparral-covered foothills, and then the pine- and fir-covered mountains as the season progresses. Following summer rains there may be a second flowering in many places, in September or even October.

For those unable to go far afield, city and county parks that retain some natural vegetation may support butterflies, but be sure that in each place collecting is permitted. Vacant lots are readily available and provide many butterfly species, particularly in summer and fall (see *The Natural History of Vacant Lots*, by Vessel and Wong, NHG no. 50).

State and national parks are off limits to collectors in general, but offer unsurpassed opportunities for butterfly watchers. Private land is now mostly tightly fenced, to prevent trespassing, especially by motorcycles and off-road vehicles. However, it is frequently possible to obtain permission to collect or observe on private lands. If such permission is given, the guest should take care to respect the owner's property, including care not to frighten livestock.

Places to look for particular butterflies will be mentioned under species accounts.

How to Prepare and Display Butterflies

Now that you have taken some butterflies, you will need to know how to preserve them. The method chosen will depend on whether you wish to display them, set them aside for future study, or retain them for exchange with other collectors. Since display requires that the butterfly's wings be spread, the simpler skills of filing them for future use will be considered first.

Labeling of specimens is the first step to good preservation. What information should be given on a specimen label? Certainly, as a minimum, the locality, date, and collector. Many labels convey insufficient information to enable another worker to locate the place accurately. What political subdivision should you use? The fifty-eight California counties are basic. Yet they are unevenly distributed as to area. Only ten form what is usually considered southern California. One, San Bernardino, if ranked as a state, would be forty-third in size: larger than Maryland but smaller than West Virginia. Within these counties, and in some cases extending beyond their boundaries, are the national forests.

Of great assistance to the collector in pinpointing collecting localities is a good map. The touring maps issued by various agencies are helpful, but one need only compare today's maps with those of a decade or two ago to appreciate that highways change, as does the system of numbering them. Much better are the topographical maps prepared by the U.S. Geological Survey, which enable one to give details up to the contour level, usually 100 feet. The national forest maps also have a grid system, and access roads and trails are marked.

In determining altitude, a good altimeter is a valuable aid. However, the commonest of these are essentially barometers as well, and must be reset to a known elevation almost daily for accurate results.

Another way to pinpoint a locality is by giving its coordinates. Township and range may be used. Latitude and longitude form another system. However, such a system requires some time to use, and has been objected to by some. Distance and direction from a known point on a map is often used. By selecting among these methods the most appropriate one for the occasion, the worker can designate localities so that someone coming even years later will most likely be able to relocate them, at least within a very short distance.

Labels are small, and limited information can be put on them without making them unduly cumbersome. However, most serious workers keep notebooks giving more information, which may include political boundaries, national forests, peaks, drainages, elevations, road miles, or air miles, and even township, range, and section, if accuracy to this detail seems desirable.

To save butterflies, store them with wings closed over the back in triangular envelopes folded as shown in Figure 14. The paper triangles may be made from pads of white paper available in several sizes from any variety store, or cut from larger sheets to the desired size. Each envelope should have written on it the name and sex of the butterfly, the locality and date where it was found, and the collector's name. These items are spoken of as minimum data. The importance of this basic information in maintaining a well-ordered collection cannot be too strongly emphasized.

Some collectors prefer to use the glassine envelopes available from stamp dealers. After the collecting information has been written on the envelope and the butterfly placed in it, the envelopes may be placed in a wooden or cardboard box, with a few crystals or flakes of naphthalene (old-fashioned moth balls) to deter mites, book lice, and carpet beetles from attacking them, and stored until needed. Specimens should not be

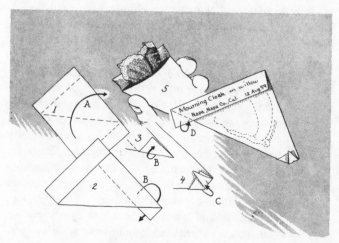

FIG. 14 Triangular envelopes

FIG. 15 Chlorocresol chamber and forceps

stored in tin boxes because changes in temperature may cause moisture to condense in the boxes, which will tend to make mold grow on the specimens.

Another method involves the use of chlorocresol and allows specimens to remain flexible for some time prior to spreading. A small plastic box with a tight lid is used (see Figure 15). A teaspoonful of crystalline chlorocresol is scattered on the bottom of the box. A layer of Cellucotton or Kleenex is laid over the chlorocresol, and fresh butterfly specimens are laid on the Cellucotton, taking care they do not overlap. When this layer of specimens is complete, another layer of Cellucotton can be placed, and more butterflies placed on it. Each layer will represent a collecting site and will carry its own label. If the same data apply to all the specimens in the box, one label on top of the box will do. When the box is full, or when all the fresh specimens have been placed in the box, the box should be sealed with tape and placed in the refrigerator. The box can be removed, opened, and the specimens spread, when needed. Chlorocresol is poisonous if taken internally and should be used carefully.

If butterflies have been allowed to dry out, they will be stiff and brittle, and will need to be relaxed before they can be spread. This requires a moistening chamber, or relaxing jar that can be made from a bell jar, tobacco jar, or almost any receptacle with a tight-fitting lid in which condensing moisture will not drop directly onto the specimens. A thick layer of absorbent material, such as sand, sawdust, or paper towels is used in the bottom of the jar. This layer is moistened with a solution made by adding a few drops of phenol to a half-pint of warm water. Use only as much liquid as the material will readily absorb. The butterflies may then be placed in a small glass or plastic tray and put in the jar, taking care that the wings do not come in contact with the liquid solution. The jar is then set in a relatively warm place in the room for overnight. Each morning test the specimens by checking their flexibility with forceps. When the wings are again flexible they may be mounted as though fresh. The lower part of a relaxing jar is seen in the background in Figure 17.

Another method of relaxing the wings is to take an empty two-pound coffee can and partly fill it with a solution of phenol as described above. Tear some clean cotton cloths into strips and immerse them into the phenol solution. After they have been thoroughly wetted, wring them out, dry the inside of the coffee can, fold the cloths, and lay these moist cloths on the bottom of the can. Put the butterflies to be mounted in a plastic or glass tray, place this tray on the moist cloths, then put the lid on the can. Set the can aside until morning, and each morning test the butterflies until they are relaxed enough to mount.

For spreading butterflies you will need one or more spreading boards, which can be made as shown in Figure 16 or purchased from a dealer. You will need several packages of insect pins, which come in several sizes, of which Number 2 is the most generally useful. You will need some pins with round glass heads, a spreading needle (which can be made by inserting the eye end of a sewing needle into the end of a matchstick), and some glassine paper cut into strips of appropriate width and length.

Take a butterfly from the relaxing jar and insert an insect pin through its thorax at right angles to the long axis of its body so

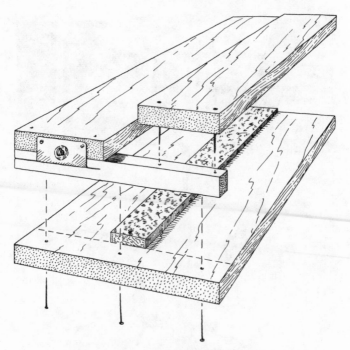

FIG. 16 Spreading board

that the pin protrudes well beyond the insect's legs. Next, pin the butterfly into the groove of the spreading board, so that its head is away from you, its abdomen lies freely in the groove, and its wings, when opened, lie flush with the surface of the board. Now take a strip of glassine paper of moderate width and long enough to cross both wings on one side of the body. Insert the pin above the upper edge of the upper butterfly wing, as shown in Figure 17, gradually depressing the wings until they lie flat. Now adjust the wings so that the upper wing has its lower edge at right angles to the long axis of the body, and the lower wing slightly underlaps the upper. Using two round-headed pins, pin the strip of paper firmly in place. Now do the same with the wings on the other side of the body. It makes no difference which side is spread first.

Now take, on each side of the insect, a strip of glassine

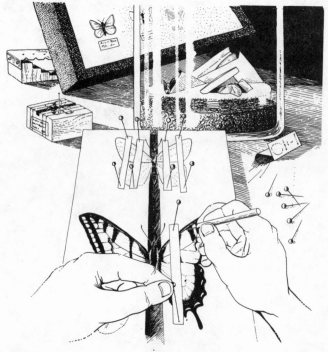

FIG. 17 Spreading butterflies

paper wide enough to cover the rest of each pair of wings, and pin down firmly with round-headed pins. Pin the antennae forward parallel to the margins of the front wings, and support the abdomen by two pins crossed underneath it.

Specimens should remain on the board until dry. The abdomen may be carefully tested with a pin. When it is thoroughly rigid, the specimen may be removed from the board. It takes fresh specimens much longer to dry than those that have been previously dried and then relaxed.

The simplest and cheapest box for storage of a study collection is a cigar box in which a pinning bottom has been placed. Cork, yucca pith, balsa wood, styrofoam, and corrugated cardboard are among the pinning surfaces that have been used or recommended. However, each falls somewhat short of being truly suitable. Pure cork is very expensive, and ground cork often is stuck together with materials that rust the pin points. Yucca pith occurs in narrow strips, and is only locally available. Balsa wood is often too hard for good pinning. Styrofoam seems very neat, but is completely eaten away by paradichlorobenzene (moth crystals, dichlorocide). Softer sorts of corrugated cardboard work well but are often hard to find and must be firmly glued to the bottom of the box. Certain of the softer of the cellulose boards are fair. But synthetic pinning bottoms made for the purpose are now readily available from dealers and work the best.

Before being transferred, the specimens should be labeled (see Figure 18). Labels may be hand lettered with a crow quill pen, or may be typed with a black ribbon on bond paper and either reduced photographically or, if a considerable number of labels is needed, by a commercial printer. Some entomologists have a hand printing press that prints in Number 4 type.

A pinning block may be used to adjust all labels to the same height. Pinning forceps, as shown in Figure 18, are convenient but not essential.

Cigar boxes are not pest proof, and should be examined frequently for infestation. The student may prefer to adopt one of

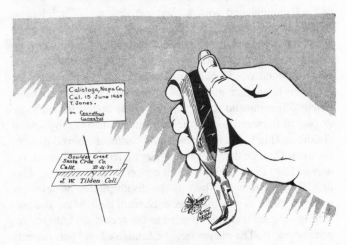

FIG. 18 Labels and pinning forceps

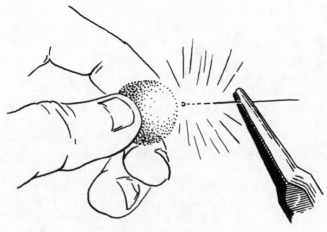

FIG. 19 Mounting moth ball on pin

the standard insect boxes obtained from dealers. Insect boxes
are supplied in several sizes and prices. Even with such boxes,
it is well to keep a fumigant inside. This may be a mothball
into which the heated head of a pin has been thrust (Figure 19),
or finely divided naphthalene flakes scattered in the bottom, or
a small container filled with paradichlorobenzene, or a strip
of tape impregnated with a chemical deterrent. Naphthalene
keeps pests from entering, and lasts quite a while, but does not
usually kill pests that are already there. Paradichlorobenzene
kills pests, but does not last as long before it vaporizes. The
student should avoid inhaling all deterrents as much as possible.

There is another method of displaying butterflies that has
much to recommend it, particularly if the butterflies are to be
displayed frequently, or if the collection is handled by school
children. This is the Riker mount, a cardboard-boxed plaque,
obtained from dealers in various sizes. Here butterflies are removed from the pins and laid on cotton behind glass (see Figure 17, background). There is the disadvantage, however, of
not being able to view the specimen from both sides. The collecting data are no longer pinned to the specimen, but may appear beside it. The plaque may be fumigated and permanently
sealed, and the specimens are safe unless the glass is broken.

As your collection grows you may wish to exchange duplicate specimens with collectors in other parts of the state or nation to broaden your knowledge. Names and addresses of other collectors may be obtained through the pages of the *Lepidopterists' News*, or the *Naturalists' Directory*. Lists of specimens desired and offered are exchanged, and specimens are sent after mutual agreement. Usually butterflies in envelopes, rather than pinned specimens, are exchanged, as these are easier to prepare for mailing and are less likely to get broken. It is as easy to relax and mount another collector's specimens as your own. Care should be taken to send only first-class specimens unless other arrangements have been made ahead of time.

History of Butterfly Collecting in California

The first naturalists to collect butterflies in California were the Russians. J. F. Eschscholtz and Adelbert Chamiso reached San Francisco in October 1816 aboard the brig *Rurik* commanded by Captain Kotzebue. They landed on Yerba Buena Island, where they found the Common Checkerspot, *Occidryas chalcedona*, as verified by H. H. Behr, who saw the specimen in the Berlin Museum. They later visited San Jose and Monterey, and Eschscholtz returned to California in 1824.

The Russians came from Siberia via the Aleutian Islands and Alaska, establishing trading posts at Fort Ross, on the coast north of the Russian River, and at Bodega Bay, in 1820. A number of Russian naturalists collected plants and insects chiefly in Sonoma County, among whom I. G. Vosnesensky is best known. There is a record of his having climbed Mt. St. Helena on June 12, 1841, the first person other than Indians known to have done so. His butterflies were sent to the Russian entomologist, Edouard Ménétriés, who named the California Dog-face *Colias wosnesenskii*. However, shortly before the name *Colias wosnesenskii* was published, in 1855, the French entomologist J. H. A. Boisduval published the description of a specimen of our dog-face collected by Lorquin as *Colias eurydice*, giving Boisduval's name priority. Our California Poppy, *Eschscholtzia californica*, was named by Chamiso for his col-

league Eschscholtz. Thus our State Butterfly and our State
Flower were very early named for Russians, even though the
butterfly name is not in current use.

The decade from 1849 to 1859 was the period following the
discovery of gold in California. Among those seeking this pre-
cious metal was the naturalist Pierre J. M. Lorquin, who arrived from France in 1849. Having previously collected insects
for Boisduval in Algeria, he continued to provide his mentor
with a steady flow of insect novelties while traveling the length
and breadth of California. Lorquin left California for the
Philippines and China in 1856, returned to California in 1860,
remained one year, and left again for China in 1862. He did not
again return to California.

The many specimens sent to Boisduval by Lorquin were described in two papers entitled "Butterflies of California." Published in 1852 and 1869, they contain descriptions of many of
our common and well-known species. The first of these is
probably the earliest paper dealing entirely with California
butterflies.

Two other events of importance to lepidopterists took place
in the 1850s: Hans Herman Behr arrived in San Francisco in
1851, where he remained until his death in 1904, and the California Academy of Sciences was founded in 1853 by a small
group of men including Behr, Lorquin, and James Behrens.

The decade from 1860 to 1870 opened with the appointment
of J. D. Whitney to the newly created office of State Geologist.
His assistants, J. G. Cooper, W. M. Gabb, and Charles Hoffmann, collected butterflies in various parts of the state. These
butterflies were described by Behr and others, and were often
given the names of their discoverers. The Sierran naturalist
John Muir similarly collected butterflies that were sent to
Henry Edwards.

The double decade from 1870 to 1890 opened with the com-
pletion of the transcontinental railroad on May 10, 1869. The
railroad made the trip from the East Coast to the West an easy
one compared to the tedious and often dangerous covered
wagon routes or the longer sea voyage around Cape Horn. If
one entomologist dominated this period in California, it was
Henry Edwards, an actor by profession, who collected widely

throughout the state and published accordingly. John Xantus de Vesey collected at Fort Tejon in the 1870s before moving to Cape San Lucas in Baja California; his moths and butterflies were sent to Behr. H. K. Morrison collected in southern California in the 1880s. It has been said that Tryon Reakirt collected in California. This statement is apparently untrue. Reakirt, a Philadelphian, named some California material that he bought from Lorquin, but there seems no proof that he ever visited California in person. O. T. Baron collected in Mendocino and Fresno counties, returning to Germany about 1890. His discoveries were described by William H. Edwards and Henry Edwards.

Other important collectors of this period were Baron Terloot de Popelaire, T. L. Mead, James Behrens, and J. J. Rivers. All but Rivers had butterflies named for them, and he himself named one.

Though not a Californian, W. H. Edwards, of Coalburgh, Virginia, shares with Boisduval the distinction of naming many California butterflies. A glance at names of the describers of North American butterflies will show that he named more of our butterflies than any one else. He obained his specimens from many others, including all of those named above, either by direct contact, or indirectly.

The activities of the California Academy of Sciences came to an abrupt halt on April 18, 1906, when its collections were destroyed by an earthquake, followed by fire. Eight boxes of insects, most or all of them types, were saved, being moved three times ahead of the fire until they found safety at Fort Mason. Alice Eastwood, botanist, assisted by Miss Hyde, librarian, and John Carleson, caretaker, effected the rescue. A number of the Academy's curators were away, cruising in the Galapagos Islands off Ecuador. Among them was F. X. Williams, a student of V. L. Kellogg at Stanford University. Williams's few butterflies and beetles, together with the reptiles and birds acquired from those inhospitable shores, became the nucleus of the Academy's future collections. Williams explored the Mt. Shasta region in 1907 (from which came a paper on the butterflies of Mt. Shasta) and the Lake Tahoe region, with Erval J. Newcomer of Palo Alto. Mt. Shasta was later to become the domain of James E. Cottle of San Francisco and Hayward. Williams, who transferred his interest to wasps, continued his work until 1967, while Newcomer continued to produce articles on Lepidoptera until his death in 1981.

Over the years, most serious lepidopterists in northern California have been affiliated with the California Academy of Sciences, and/or with the closely associated Pacific Coast Entomological Society, established in 1901. Early curators of the Academy were Hans Herman Behr (1862–1868 and 1880–1904); Richard Stretch (1868–1880), whose primary interest was in Lepidoptera; E. C. Van Dyke (1904–1916); and E. P. Van Duzee (1916–1940), whose primary interests were Coleoptera and Hemiptera, respectively, but who provided encouragement for those interested in butterflies, since both had been butterfly students in their early years. The *Pan-Pacific Entomologist*, a periodical established in 1924, provided for publication of original research for entomologists, and many articles on butterflies have appeared in its pages.

The period of the mid-1920s to early 1940s was a time of considerable activity among northern California lepidopterists. In addition to those whose work centered around the California Academy of Sciences, a number worked out of the Entomology Department of the University of California, Berkeley, and others out of the Biology Department at San Jose State College (now University).

Many were associated in one way or another with Robert G. "Bob" Wind and his Pacific Coast Biological Service in Berkeley, whose home, and later, store, became their informal meeting place. Among those active during this period were the Bohart brothers, Richard and George; Thomas Davies; E. A. Dodge; Michael Doudoroff; Graham Heid; William Hovanitz; W. Harry Lange; George Mansfield; the Smith brothers, Art and Ed; R. F. Sternitzky; J. W. Tilden; and, of course, Bob Wind. Some have died, others went on to other interests in later years, but some are still active today.

One butterfly collector active in the Berkeley area in the late 1920s is one of our best-known conservationists today—longtime Executive Director of the Sierra Club and founder of Friends of the Earth, David R. Brower.

For some years, Stanford University has been an active center for research on Lepidoptera under the guidance of Paul Ehrlich. Jerry Powell and his many students at the University of California, Berkeley, have contributed greatly to our knowledge of butterflies. Arthur Shapiro, at the University of California, Davis, has carried on and supervised many studies on butterflies. Others active in the San Francisco Bay region are Robert Langston, John Lane, R. M. Brown, and Larry Orsak, whose work on endangered species is extensive. These are but a few of many current workers.

At the Santa Barbara Museum of Natural History, Scott E. Miller has replaced an earlier collection destroyed by fire with a regional collection emphasizing the California Channel Islands. Paul Opler, of San Jose State University and the University of California, formerly with the federal Endangered Species Program, is now editor of the *Bulletin of the Entomological Society of America*.

A pioneer lepidopterist of southern California, whose activities spanned the late 1800s and early 1900s, was William Greenwood Wright, who came to San Bernardino in 1873 and operated a planing mill there until 1897, when he retired to devote himself fully to butterfly collecting. Wright spent fifteen years gathering material for a book, traveling from Alaska to Mazatlan, Mexico. Although replete with errors, his *West Coast Butterflies*, published in 1905, is a remarkable accomplishment for its time, and copies are collector's items. The collection of specimens on which his book is based was given to the California Academy of Sciences.

In 1909, the activities of a group of lepidopterists centered in Pasadena began to be chronicled in the *Pomona College Journal of Entomology and Zoology*, Claremont. In that year Victor L. Clemence summered in Europe, visiting museums in London and Tring (the Rothschild collection), and in Rennes, France (the Oberthur collection). In 1920, he made a trip to the Huachuca Mountains of Arizona with Karl R. Coolidge of Palo Alto, who that year moved to Pasadena and became secretary of the Entomological Club. Fordyce Grinnell, Jr., club president, compiled material for a book on pioneer naturalists of California, and showed interest in the mimetic relationship of Lorquin's Admiral (*Basilarchia lorquini*) and the California Sister (*Adelpha bredowii californica*).

Wilhelm Schrader of Los Angeles built a special laboratory for his breeding experiments on the Gulf Fritillary (*Agraulis vanillae incarnata*), the Buckeye (*Junonia coenia*), and the common Checkerspot (*Occidryas chalcedona*), among others. J. J. Rivers of Santa Monica compiled notes on the checker-spots (at that time all in genus *Melitaea*, now in several genera). Leo Goeppinger found the San Emigdio Blue (*Plebulina emigdionis*) and Becker's White (*Pontia beckerii*) in Kern and Inyo counties. G. R. Pilate collected in the Kern River country for William Barnes, a Decatur, Illinois, physician. Barnes, alone and with coauthors Lindsey, McDunnough, and Benjamin, described many butterflies, some from California. Don Rose of Pasadena made an early collecting trip to San Clemente Island.

Among scientists whose earliest exploits were with butter-flies were Alexander Agassiz, son of the great Louis Agassiz, who collected Lepidoptera in northern California and wrote his first scientific paper on the flight of butterflies and moths; V. L. Kellogg of Stanford University, who wrote on butterflies of mountain summits (an early hilltopping study) for the *Sierra Club Bulletin* of July 1913; and E. O. Essig of the University of California, Berkeley, who published an article on *Vanessa* while serving as horticultural commissioner of Ventura County.

Butterflies lost but rediscovered were the Veined Blue (*Icaricia neurona*), described by Henry Skinner from a single fe-male in 1902, rediscovered by Hal Newcomb, formerly of Boston, where he was president of the Cambridge Entomologi-cal Society, but later of Pasadena (South Pasadena), Karl R. Coolidge, and J. R. Haskin of Los Angeles on Mt. Wilson in the San Gabriel Mountains in September 1912; and Edwards's Swallowtail (*Papilio indra pergamus*), described by Henry Ed-wards from Santa Barbara in 1875 (but not since taken there), rediscovered by Janet Riddell and Margaret Fountaine near the Arrowhead landmark above San Bernardino, probably during World War I (1914–1918).

Butterfly farming, or rearing of Lepidoptera for pleasure and profit, had an early exponent in Ximena McGlashan, who with her father, Charles F. McGlashan, operated a butterfly farm in Truckee, California, between 1912 and 1914. They also published *The Butterfly Farmer, a Monthly Magazine for Amateur Entomologists*, which achieved a moderate circulation. A later exponent of butterfly farming was Albert Carter of Roscoe, California. Active in the late 1920s and early 1930s, he published the *Butterfly Park Nature Club News*, of which four volumes appeared between March 1929 and January 1932. A young lepidopterist who worked for him was Lloyd M. Martin.

In Los Angeles, the Southwest Museum was host to a group of enthusiasts organized in 1915 by Fordyce Grinnell, Jr., brother of Joseph Grinnell of the University of California Museum of Vertebrate Zoology. Called the Lorquin Club, its members included those interested in reptiles and shells as well as insects. Between August 1916 and January 1919 the club published *Lorquinia*, of which two volumes containing botanical as well as entomological papers and notes appeared.

In 1927 this group changed its name to the Lorquin Entomological Society, and its affiliation to the Los Angeles County Museum under the guidance of John A. Comstock, whose book *Butterflies of California* appeared that year. A month-long Butterfly Show was an annual event.

Members participating in this and other activities during the late 1920s and early 1930s included C. M. Dammers, F. W. Friday, John S. Garth, J. D. Gunder, Charles Ingham, George Malcolm, Don Meadows, Hal Newcomb, C. N. Rudkin, and John Sperry.

Under the guidance of Comstock and Dammers, the earlier emphasis on collecting changed to life history studies. Students who became interested in this aspect and went on to formal studies in butterfly ecology and population genetics included John and Thomas Emmel, Chris Henne, Noel McFarland, and Rudolph Mattoni. The potential for using butterflies as subjects for scientific research was realized in 1964 with the establishment by the late William Hovanitz of Arcadia, of the *Journal of Research on the Lepidoptera*, now in its twenty-third volume as of this writing (1984).

Outstanding contributions to lepidopterology by William

Hovanitz were many. He demonstrated how environment and heredity combine to produce color patterns seen in butterflies such as California's Common Checkerspot (*Occidryas chalcedona*); how hybridization renews variability in butterflies such as the sulfurs *Colias eurytheme* and *philodice*; he did extensive studies worldwide on the difficult *Colias* butterflies; he showed how butterfly distribution coordinates with latitude and altitude in North and South America; and he encouraged his students and colleagues to make contributions. His untimely death left a great void.

Entomologists living in southern California have looked to the Natural History Museum of Los Angeles County and the associated Lorquin Entomological Society for guidance and support. Curators of entomology and their primary interests were Lyman J. Muchmore, Coleoptera (1923–1936); John Adams Comstock, Lepidoptera (1927–1948); Lloyd M. Martin, Lepidoptera (1936–1969); William Dwight Pierce, Coleoptera (1937–1951); Christopher Henne, Lepidoptera (1940–1946); E. Graywood Smyth, Coleoptera (1946–1951); Fred S. Truxal, Hemiptera (1952–1961); Charles L. Hogue, Diptera (1962–); Roy R. Snelling, Hymenoptera (1963–); and Julian P. Donahue, Lepidoptera (1970–). Edited by Comstock, the *Bulletin of the Southern California Academy of Sciences* published many early descriptive and life history studies written by lepidopterists of the region.

Emphasizing insects of San Diego County, the southwestern United States, and Baja California, the San Diego Natural History Museum collection has been curated by the following entomologists, with specialties and dates of tenure: G. H. Field, Coleoptera and Lepidoptera (1921–1923); W. S. Wright (not to be confused with W. G. Wright, whose book *West Coast Butterflies* appeared earlier), Lepidoptera (1923–1933); I. Moore, Coleoptera (1933–1934); C. F. Harbison, Lepidoptera (1935–1969); F. T. Thorne, Lepidoptera (1972–1975); D. K. Faulkner, Lepidoptera and Neuroptera (1975–), with J. W. Brown, Lepidoptera (1977–), as assistant. In addition, F. X. Williams, Lepidoptera and Hymenoptera (1953–1968), and J. A. Comstock, Lepidoptera (1955–1970), were honorary curators of entomology. Under their successive guidance a group known as the Entomology Club met monthly at the museum (1937–1943). Reorganized as Los Entomologos ("the entomologists") after World War II, during which the museum was used as a navy hospital, the group has been active intermittently (1953–1955; 1962–1965, when a newsletter was published; and 1979–). Among lepidopterists developed by this group were Robert Langston, Jerry A. Powell, Oakley Shields, and Ray Stanford.

The history of butterfly collecting and of research on the Lepidoptera of California is a long and distinguished one. Its final chapters, however, remain to be written. The purpose of this book will be achieved only when you, the reader, are inspired to add to it your own observations on California butterflies and the conclusions that you have drawn from them.

Variation and Hybridization in Butterflies

Class Variation in Butterflies

As we examine our butterflies, we may notice that not all individuals of a given kind are alike. They may differ among themselves in one of two ways—either as individuals, or as classes of individuals. If the variation is so slight as to be barely noticeable when the specimens are arranged in a series, it is said to be continuous variation. If the variation is abrupt, so that the individuals fall into two or more clear-cut categories, it is called discontinuous variation or polymorphism (*poly* = "many"; *morph* = "form").

The commonest kind of polymorphism occurs when males and females of the same species differ from one another in color or markings. This kind of variation is called dimorphism ("two forms"). Such a species is said to be sexually dimorphic. While all fritillaries are dimorphic to some degree, in two of our California species the difference is striking. In the Leto Fritillary (*Speyeria cybele leto*), and the Apache Fritillary (*S. nokomis apacheana*), the males are reddish as fritillaries usually are, but the females are dark, with light borders. Again, while the silhouette of the dog's head appears on the forewings of the male California Dog-face, the forewings of the female are entirely yellow except for the dark "eye."

In some species, one sex may have more than a single form.

Among the sulfurs of the genus *Colias*, this sex is the female, which occurs in the normal yellow form, or in a white (albinic) form. The males are always of the yellow form. This dimorphism may be observed in any alfalfa field where the Alfalfa Butterfly flies.

In the California Dog-face, dimorphic males as well as females occur. Most males have the hind wing all yellow, but some males have the hind wing black margined. This variation has been named form "*bernardino*." Some females have the front wing with a variable shading of light brown. This variation has been termed form "*amorphae*." It should be noted that whereas form names are convenient to use as names to separate different-looking individuals, they have no taxonomic standing, and are not an integral part of a scientific name. For this reason they appear in quotation marks in the checklist and text, although not in the index.

Another sort of polymorphism occurs when succeeding broods of a species differ in recognizable ways. The first brood of the Sara Orange-tip (*Anthocharis sara*) to emerge in the spring is smaller and darker than succeeding broods, and has been called Reakirt's Orange-tip, form "*reakirtii*." Again, this name has no taxonomic standing but has been in general use for so long that it has become familiar. Summer and fall broods of the Alfalfa Butterfly (*Colias eurytheme*) are often quite different in appearance, the summer brood, form "*amphidusa*," being a brilliant orange yellow, whereas the fall brood, "*autumnalis*," is much smaller and a much duller yellow. The Mustard White (*Artogeia napi*) has well-marked brood forms, which have in some cases received names. Such variation follows an annual cycle and is known as seasonal polymorphism.

Continuous variation is less easy to observe. The Common Checkerspot (*Occidryas chalcedona*) occurs widely in California, although there are large areas where it is not found. Some of its populations intergrade imperceptibly, tending to be darker in cool, moist climates and lighter or more reddish in warm, dry ones. In some cases it may be difficult to distinguish between named populations occupying adjacent localities. But in other cases, adjacent populations are quite different in appearance. Usually little difficulty is encountered in distinguishing between populations that occur in widely separated localities.

Individual Variation in Butterflies

Anyone who collects butterflies is likely to encounter a West Coast Lady (*Vanessa annabella*) with white spots in place of the blue ones normally found on the upper side of the hind wing, or a Buckeye (*Junonia coenia*) in which the peacock spots are enlarged or even fused. When an individual butterfly differs markedly from the norm for the species, it is said to be aberrant (*ab + errans* = "wandering from"), and is called an aberration, a better term than *sport* or *freak*. At one time aberrations were named, but zoologists came to an agreement that scientific names should be applied to populations, not to individuals, and so the practice was wisely discontinued.

Among many kinds of butterflies, such as checkerspots and fritillaries, there is often a tendency toward blackness, or melanism, rather than whiteness, or albinism. It was found by experimental breeding that such anomalies could be produced by abruptly changing the temperature and humidity at certain stages in the life cycle of the butterfly. An aberrant Gulf Fritillary showing pronounced melanism is shown in Figure 20.

A sometimes startling form of individual variation is the

FIG. 20 Aberrant Gulf Fritillary

♀   ♂

FIG. 21 Bilateral gynandromorph of Purplish Copper

hermaphrodite or gynandromorph (the preferred term), in which both male and female characters are present in the same individual. This form of variation is most easily apparent in species that are sexually dimorphic, the male and female being quite different in appearance. Such variation may involve only a small part of one wing, or involve a large share of the insect. In cases in which the distribution of these sexual variations is patchwork, the individual is called a mosaic. A striking type of gynandromorph is the bilateral, in which one entire side of an insect is one sex, the other side, the other sex. A bilateral gynandromorph of the Purplish Copper (*Epidemia helloides*) is shown in Figure 21.

Individual variation occurs very commonly in the populations of some species and subspecies. All populations of the Leanira Checkerspot (*Thessalia leanira*) tend to individual variation, but in certain places, such as near Wrightwood, on the northern side of the San Gabriel Mountains, between mountains and desert, a series may contain individuals suggesting subspecies *wrightii, cerrita*, and *alma*. And populations of the Crocale Patch (*Chlosyne lacinia crocale*) seldom have two identical individuals. Those with a rufous band across the wings have been called "*rufescens*"; those with a white band, "*crocale*"; and those black with little or no band, "*nigrescens*." All manner of intermediates occur. C. M. Dammers has shown that all the varieties can be produced from a single mating.

Hybridization in Butterflies

The usual criterion for a valid species is that it does not normally interbreed with other species. Most butterflies follow this rule most of the time, but there are several groups of species that will occasionally interbreed in nature, on the meeting line between these populations. One such group is the admirals, genus *Basilarchia*. The Mono Lake Basin, east of the Sierra Nevada, is such a place. Here the ranges of *B. lorquini* and *B. weidemeyerii latifascia* overlap. On the willows that surround Mono Lake, both species may be seen, together with an occasional hybrid. These hybrids were named *fridayi* by Gunder after their discoverer, F. W. Friday, and are still spoken of as Friday's Admiral. (There seems to be no reason not to use common names for repetitive hybrids, such as many of our cultivated flower varieties.) But today scientific names of hybrids have no official standing, so the correct designation is *B. lorquini* × *B. weidemeyerii latifascia*. It should be remembered that these hybrids are neither species nor subspecies.

A similar situation exists in Oak Creek Canyon, Arizona, where the ranges of the Rocky Mountain *Basilarchia weidemeyerii angustifascia* and the Arizona subspecies of the Red-spotted Purple meet. Here also, both species are found, together with an occasional hybrid showing characteristics of both. Gunder named this hybrid *doudoroffi* after the man who first found it. Then, in 1934, it was usual to name aberrations and hybrids. This is no longer the case. So, again, in speaking of a hybrid, we would use the combination *B. weidemeyerii angustifascia* × *B. arthemis arizonensis*. Hybrids of other species and subspecies of *Basilarchia* are found in other parts of North America.

Other butterflies known to hybridize are the swallowtails, genus *Papilio*. In California, specimens of the Two-tailed Swallowtail (*Papilio multicaudatus*) taken in the Tehachapi Mountains of Kern County suggest hybridization with the Western Tiger Swallowtail (*Papilio rutulus*). The latter is sympatric (*sym* = "same"; *patria* = "country") with the Pale Swallowtail (*Papilio eurymedon*) over much of their ranges, yet hybridization between these two species is almost unknown.

A remarkable case is that of the orange-tip named *Antho-charis dammersi* by Ingham in 1933. Often called Dammers's Orange-tip, it resembles the Sara Orange-tip (*A. sara*) on the upper side and Grinnell's Marble (*Falcapica* [formerly *Anthocharis*] *lanceolata australis*) on the under side. Originally described as a full species, and long so regarded, it is probably a hybrid between these two species, now considered to belong in different genera. Only three specimens have ever been taken.

How to Tell a Butterfly from a Moth

That moths and butterflies are closely related is readily apparent. People have noted that the wings of a butterfly or moth are covered with tiny scales that cling to the fingers if the wing is touched or rubbed carelessly. Both belong to the order Lepidoptera ("scaly wings"), a characteristic that sets them apart from all other insects (see Figure 22).

While it is true that most butterflies fly in the daytime, it is not true that moths fly only at night. Many moths are day-fliers. And whereas some moths spin silken cocoons, some others

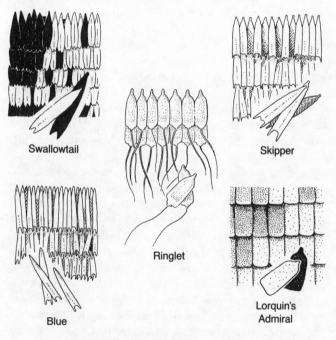

Swallowtail

Skipper

Ringlet

Blue

Lorquin's Admiral

FIG. 22 Types of butterfly wing scales

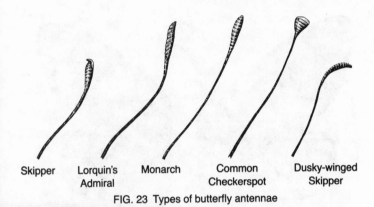

Skipper Lorquin's Monarch Common Dusky-winged
Admiral Checkerspot Skipper

FIG. 23 Types of butterfly antennae

have pupal cases as naked as a butterfly chrysalis, and skippers usually spin a slight cocoon. Again, whereas some moths are heavy bodied, others are as slender bodied as a butterfly. And whereas some moths are plain colored, many are as bright as a brightly colored butterfly, and many butterflies are quite plain.

Fortunately, the antennae, or feelers, provide an easy way to distinguish between moths and butterflies. The antennae of moths may be featherlike, comblike, pointed, or very slender, so moths are said to be heterocerous (having different kinds of antennae). The antennae of butterflies are always club-shaped, enlarged at the tip, a condition said to be rhopalocerous (having clubbed antennae). For convenience, moths are often spoken of as the Heterocera, and butterflies as the Rhopalocera. Figure 23 shows some of the types of antennae found in butterflies.

Day-flying Moths

Whereas few butterflies fly at night (although some tropical species fly just before sunrise, and in the evening twilight), there are many moths that fly in the daytime. The less experienced butterfly student will encounter them, and may be puzzled when trying to find them in a butterfly book. If the antennae are observed, however, it will be noted that they are not clubbed, a sure clue that they are moths, not butterflies.

Garth remembers a day on Mt. Shasta when the commonest insects in flight were salmon pink saturniid moths the size of

FIG. 24 Sheep Moth

fritillaries. Called Sheep Moths (*Hemileuca eglanterina shastaensis*), they were present by the hundreds, and in infinite variety, from yellow to red, and from cream color to nearly black. The related Nevada Buck Moth (*Hemileuca nevadensis*) flies over the sagebrush flats east of the Sierra Nevada and is found in the San Joaquin Valley and near Corona in southern California as well. The spiny larvae of these day-flying moths may be gathered for rearing. Do not pick them up by hand; their spiny hairs sting severely. The larval food plants of the Sheep Moth (Figure 24) are very diverse. Willow, snowberry, Bitterbrush, and Buck Brush are common foods. The Nevada Buck Moth prefers willow or cottonwood.

At midelevations in our mountains, black-and-white pericopid moths (*Gnophaela latipennis*) and so-called wasp moths of the genus *Ctenucha*, family Amatidae, may be found. *C. brunnea* has wings that are quite brownish. *C. multifaria* and *C. rubroscapus* have very dark to black wings, iridescent bluish bodies, and red shoulder lappets, and are quite handsome.

Small clear-winged hawk moths of the genus *Hemaris* are heavy bodied, the wings transparent except for a dark border. These resemble bumblebees as they flit from flower to flower, uncoiling their long tongues to probe deeply into tubular flowers. The clear-winged hawk moths are smaller but in habit resemble others of their family. The commonest of these, the White-lined Sphinx (*Hyles lineata*) (formerly *Celerio*) may be seen by day, especially in early morning or late afternoon, wherever it is common, visiting flowers. It is partial to evening primroses, petunias, and many flowers with deep corollas.

The early spring flowers of the serpentine hills, and of the wildflower-carpeted foothills and valleys, as well as the high rocky slopes of the Sierra Nevada, all harbor brightly colored noctuid moths of several genera. Those of the genus *Synedoida* are among the showiest. With their bright orange or reddish underwings they might easily suggest butterflies.

Thus at all elevations, and in both montane and desert habitats, one should expect to encounter day-flying moths as well as butterflies and must look for their identification in books devoted to moths, to all Lepidoptera, or, for common species, to all insects.

3 · BUTTERFLY IDENTIFICATION

To look up information about a butterfly, to discuss it with others, or to exchange it with another student, you will need to know its name. The first step is to know the family to which it belongs. Some families have easily seen characters, such as swallowtails with their long tails, nymphalids with their reduced front legs, and metalmarks with their extra-long antennae. The characters of some families are less easy to see.

In some families it is simple to tell the sexes apart, the wing patterns and colors of the males and females being quite different, as in many lycaenids, such as the blues. In other families, the sexes are very similar in appearance. In such cases, structure can be used to separate the sexes. Males usually have a rather slender abdomen, ending often in a tuft of hairs, which conceals a lateral pair of valves (harpes). In some, such as the swallowtails and pierids, the valves are easily visible. The abdomen of the female is usually much more enlarged, and pointed at the tip. Close examination shows a crosswise slit, the sex opening.

Insects, including butterflies, have few to many thickened, usually tubular, veins in their wings. The course of these veins is extremely important in classification. The main structures of a butterfly, and the wing veins, are shown and named in Figure 25. These structures should be reviewed before using the key on page 64.

The sex organs (genitalia) of insects, including butterflies, provide characters of great value in separating species, and are much used by advanced students. Such information may be

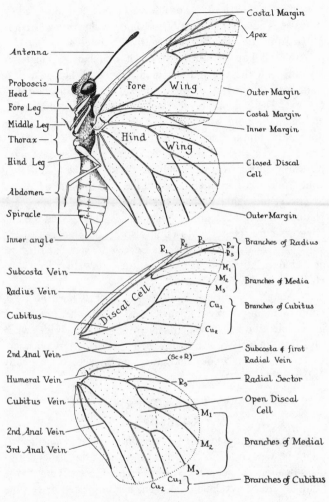

FIG. 25 Adult structures

found in more specialized literature. If a student wishes to look at the valves (harpes) of a male, these may be seen by carefully brushing away the scales from the tip of the abdomen. A few references to the male valves are mentioned in this book.

Structural characters are used in the key to the families of butterflies and skippers. (For information on how to use a key, if its use is not obvious, see *Introduction to the Natural History of the San Francisco Bay Region*, by Smith, NHG no. 1).

After finding the family to which your butterfly belongs, leaf through the illustrations in this book. When you have found one that matches yours, confirm the identification with the text. There should be correspondence in size, color, habit, and habitat before you conclude they are the same. If still in doubt, consult other sources of information, such as the References given at the end of this book. Or take your specimen to a natural history museum. Usually a curator will be glad to help you. Collections are maintained at the California Academy of Sciences, the Santa Barbara Museum of Natural History, the Natural History Museum of Los Angeles County, and the San Diego Natural History Museum; also at the branches of the University of California, such as Berkeley, Davis, and Irvine; and at the California State Universities, including San Jose, Hayward, and San Francisco.

Although it is true that forms and hybrids have no taxonomic standing at present, there was a time when they were given scientific names. Such named forms and hybrids, when illustrated in well-known books such as Holland or Comstock, are noted in this book, and, for convenience, occasionally illustrated.

How Living Things Are Classified

Before proceeding to use the classification of butterflies, it might be well to learn, or at least to review, the principles of classification in general. The science of classifying living things is called systematics. Organisms are placed first in the most inclusive groups, those that include large numbers of organisms, then progressively into smaller groups that include fewer organisms. The groups into which organisms are placed are called categories. The differences in structure, behavior, habits, and other criteria useful in placing organisms where they belong are called characters. Some characters are considered more basic for this purpose than are others.

Organisms that share many basic characters are placed in a

more inclusive category; those that share fewer or less basic characters are placed in a less inclusive one. All insects have six legs, but only Lepidoptera have scaly wings. On the basis of successively less inclusive characters, organisms are split into smaller and smaller groups until the basic, or unitary group, is reached. This is, by consensus, the species, although subdivisions may be carried further, to population level.

It is believed that closely similar organisms are related and share a more or less common ancestry. Classification based on what are believed to be natural evolutionary relationships is said to be phylogenetic.

The science of naming organisms is called taxonomy, and the grammatical rules that govern taxonomy form what is called nomenclature.

A name that is given has several parts, each of which has a special meaning. If we take the name of the Monarch, it reads like this:

<u>Danaus plexippus</u> (Linnaeus) 1758

The first name is the generic name and is capitalized. The second name is the specific name and is not capitalized. The underlining of the name indicates that it will appear in italics when presented in print. The name Linnaeus is that of the man who first published a scientific description of this species. The name of Linnaeus, the describer (author), is in parentheses to indicate that the specific name is now in a different genus than that in which it was originally described. In less formal citations, the author and date of description are usually omitted.

If we take the classification of the Monarch, and list the successive categories from the most inclusive to the least inclusive, the major categories appear like this:

Kingdom Animalia (animals)
  Phylum Arthropoda (animals with jointed feet)
    Class Insecta (insects)
      Order Lepidoptera (scaly-winged insects)
        Family Danaidae (milkweed butterflies)
          Genus *Danaus* (monarchs)
            Species *plexippus* (the Monarch)

The species, however, is the only category (or taxon, as each level is called) that has an objective existence in nature.

Butterfly Name Changes

Those who return to the study of butterflies after a lapse of some years are surprised to discover that, while they remember the butterflies themselves well enough, many of the names have changed beyond recognition. Older persons, now retired, who would like to resume an avocation that gave them pleasure in their youth, may find it difficult to master the new terminology (nomenclature), and so may become discouraged. This is both unfortunate and unnecessary.

Scientific names are changed, not capriciously, but in accordance with rules set down by a worldwide organization of zoologists, the International Commission on Zoological Nomenclature. The rules are acted upon, and changes in the rules are made, at international meetings. The name changes we see are made within the limits of the rules, or they are not recognized by other workers.

The most frequent reason for changing a name is the finding of an older name that may be correctly applied to the genus or species. The older name takes precedence (has priority) over the more recent (junior) name, which is suppressed as a synonym (= two names for the same thing). A second reason for name changing is finding that two species formerly thought to be distinct are either identical or related at the subspecies level. In the former case the junior name is suppressed as a synonym in favor of the older (senior) name. If the junior name proves to be that of a subspecies of the senior, it is indicated by a second name, forming a total of three names, spoken of as a trinomial. An example: *Vanessa atalanta rubria*, the Red Admiral.

Another reason for name changing arises when an entire butterfly assemblage (fauna) of one region, say North America, is compared with that of another region, say Europe. If similar butterflies with different generic names are found to belong to the same genus, the junior generic name becomes a synonym of the senior. If similar butterflies with different specific names are found to belong to the same species, the junior name becomes a synonym or is reduced from a full species to a subspecies.

Yet another reason for name changing occurs when a large genus, with many species, is considered by a worker on the group to consist of more than one genus. Under the existing rules governing systematics, he or she is free to divide (or split) the existing genus into two or more genera, retaining the first described species, and possibly others, in the original genus, and placing one or more species in each of the new genera that has been established (erected). Such an action has taken place lately (1977) in the genus *Philotes*, which formerly held a number of species. Of these, one now remains in *Philotes*. One other has been removed to the new genus *Philotiella*. The rest are now placed in another new genus, *Euphilotes*.

Lastly, when certain species are carefully studied, it sometimes is found that they actually consist of more than one species. In such cases, the one most similar to the original description, or matching the type specimen, retains the older name. The other, being now unnamed, is given a new name in a description that must follow certain rules. Or a species may be found to consist of two or more dissimilar populations with different ranges. These populations may be described as subspecies, one of which (the older) always bears the species name. Example: *Euphilotes enoptes enoptes*.

Let us see how the Law of Priority applies to name changes of some California butterflies. Not long ago it was discovered that our West Coast Lady, long known as *Vanessa carye*, was not entitled to the name *carye*, which was first given to a species found in Chile. The name *carye* was a homonym (= two things with the same name), and had to be changed for the younger of the two species to which it had been applied. Thus our North American West Coast Lady is now correctly known as *Vanessa annabella*.

There is a procedure whereby names in long use may be preserved (conserved) for posterity. This procedure was followed when it was found that the familiar *Danaus plexippus* was thought to be a junior name to the less familiar *menippe*. The International Commission on Zoological Nomenclature (ICZN for short), which has the power to set aside rules, intervened when requested (petitioned), suspending the Law of Priority and thus allowing the name *Danaus plexippus* to prevail.

What should become of a common name that is based on a scientific name that changes? The Mountain Vagabond was a perfect common name for the fritillary *Argynnis* (now *Speyeria*) *montivaga*. But because *montivaga* has been shown to be a synonym of *egleis*, its common name has been changed in the present field guide to the Egleis Fritillary. Similarly, Rudkin's Swallowtail was the common name of the Colorado Desert butterfly long known as *Papilio rudkini*. But it was found that *rudkini* was a subspecies of the widely ranging *Papilio polyxenes* and a synonym of the earlier described *P. coloro* Wright. Its common name has been changed accordingly from Rudkin's Swallowtail to Wright's Swallowtail. The Central Californian subspecies of *Glaucopsyche lygdamus*, the Silver Blue, was long believed to be *behrii*. Later it was found that the name *behrii* actually applied to a form of the Xerces Blue, leaving the *lygdamus* subspecies without a name. But when this subspecies was given the name *incognitus* (unrecognized), it seemed best to continue to use the familiar common name, Behr's Blue.

So, to all of our readers who take up the study of butterflies anew, don't be discouraged by name changes, which are bound to continue. Instead, look for the reasons behind these changes. If you do so, you will find the study of butterfly names a fascinating subject in itself, and an interesting aspect of natural history.

In the species accounts that follow, where generic changes have been made recently, the generic name in use before the present one is given in parentheses. Example: *Pontia occidentalis* (formerly *Pieris*).

Key to the Butterfly Families of California

1a. Antennae not ending in a club . . . . Moths (p. 55)
1b. Antennae ending in a club . . . . . . Butterflies. See #2
2a. Antennae arising far apart on head; basal segment of antenna with a hair tuft ("eye lash"); antennal club usually

ending in a point (apiculus); all veins of forewing simple, unbranched .... Skippers. See #3

2b. Antennae arising close together on head; basal antennal segment without "eye lash"; antennal club simple; some veins of forewing branched .. True Butterflies. See #4

3a. Head not wider than thorax; antennae without a well-developed apiculus; hind tibia with only one pair of spurs; large, robust insects .... Megathymidae (p. 151)

3b. Head as wide as, or wider than, thorax; antennae with an apiculus; hind tibia with two pairs of spurs ...... Hesperiidae (p. 153)

4a. Palpi very long, directed straight forward .... Libytheidae (p. 116)

4b. Palpi of normal length .... See #5

5a. Front legs of both sexes reduced in size, not used for walking, carried folded close to body .... See #6

5b. Front legs of females of normal size; front legs of males reduced in some families .... See #8

6a. Some veins of forewings swollen at base; dull-colored butterflies with eyespots .... Satyridae (p. 68)

6b. No veins of forewing swollen at base .... See #7

7a. Antennae without scales on upper surface; male with a black scent-pouch on hind wing; large orange brown or purplish brown butterflies with black wing veins ...... Danaidae (p. 71)

7b. Antennae scaled; males without a scent pouch; cell of hind wing open; small to large; often brightly colored .... Nymphalidae (p. 73) and Heliconiidae (p. 73)

8a. Antennae curved; hind wing with only one anal vein; large species, often with hind wing tailed .... Papilionidae (p. 96)

8b. Antennae straight; hind wing tails, if present, very small and slender; medium-sized to small species ..... See #9

9a. Base of antennae not touching eyes; tarsal claws bifid; medium sized; white or yellow with dark markings .... Pieridae (p. 103)

9b. Base of antennae touching eyes; small butterflies ..... See #10

10a. Antennae ⅔ as long as anterior margin of forewing; front legs of male not spined, less than ½ size of middle legs; hind wing with humeral vein ..... Riodinidae (p. 117)

10b. Antennae shorter, about ½ as long as anterior margin of forewing; front legs of male spined, more than ½ size of middle legs; hind wing without humeral vein; blue, coppery, brown, or gray; some with tiny hind wing tails .... Lycaenidae (p. 119)

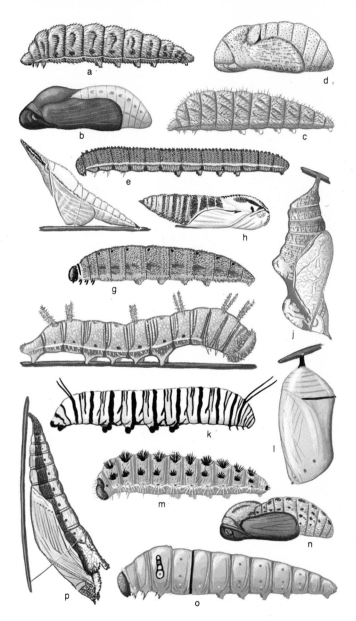

PLATE 1. **a,** larva, **b,** pupa of Acmon Blue, *Icaricia acmon* (p. 143); **c,** larva, **d,** pupa of Common Hairstreak, *Strymon melinus pudicus* (p. 131); **e,** larva, **f,** pupa of Reakirt's Orange-tip, *Anthocharis sara reakirtii* (p. 113); **i,** larva, **j,** pupa of California Sister, *Heterochroa bredowii californica* (p. 95); **k,** larva, **l,** pupa of Monarch, *Danaus plexippus* (p. 72); **m,** larva, **n,** pupa of Behr's Metalmark, *Apodemia mormo virgulti* (p. 117); **o,** larva, **p,** pupa (inverted) of Western Tiger Swallowtail, *Papilio rutulus* (p. 101). From paintings by C. M. Dammers, courtesy of Los Angeles County Natural History Museum.

Note: All plate figures slightly reduced.

PLATE 2. **a,** Woodland Satyr, *Cercyonis sthenele silvestris* (p. 70); **b,** Little Satyr, *C. s. paula* (p. 70); **c,** Least Satyr, *C. oeta* (p. 70); **d,** Ox-eyed Satyr, *C. pegala boopis* (p. 69); **e,** Baron's Satyr, *C. p. boopis* f. *baroni* (p. 69); **f,** Hoary Satyr, *C. p. boopis* f. *incana* (p. 69); **g,** Stephens's Satyr, *C. p. ariane* f. *stephensi* (p. 69); **h,** Ivallda Arctic, *Oeneis, ivallda* (p. 71); **i,** Riding's Satyr, *Neominois ridingsii* (p. 69); **j,** Chryxus Arctic, *Oeneis chryxus stanislaus* (p. 71); **k,** Iduna Arctic, *O. nevadensis iduna* (p. 70); **l,** Great Arctic, *O. n. nevadensis* (p. 70).

PLATE 3. **a,** Monarch, *Danaus plexippus* (p. 72); **b,** Striated Queen, *D. gilippus strigosus* (p. 72); **c,** Arizona Viceroy, *Basilarchia archippus obsoleta* (p. 94); **d,** Snout Butterfly, *Libytheana bachmanii larvata* (p. 116); **e,** Variegated Fritillary, *Euptoieta claudia* (p. 73); **f,** Gulf Fritillary, *Agraulis vanillae incarnata* (p. 73); **g,** California Sister, *Adelpha bredowii californica* (p. 95); **h,** Ringless Ringlet, *Coenonympha ampelos elko* (p. 68); **i,** California Ringlet, *C. california* (p. 68); **j,** Ochraceous Ringlet, *C. ochracea* (p. 68).

PLATE 4. **a,** Western Meadow Fritillary, *Clossiana e. epithore* (p. 78); **b,** Apache Fritillary, *Speyeria nokomis apacheana* (p. 74); **c,** Leto Fritillary, *S. cybele leto* (p. 74); **d,** Hydaspe Fritillary, *S. hydaspe* (p. 78); **e,** Common Checkerspot, *Occidryas c. chalcedona* (p. 79); **f,** Dwinelle's Checkerspot, *O. c. dwinellei* (p. 79); **g,** Olancha Checkerspot, *O. c. olancha* (p. 79); **h,** Sierra Checkerspot, *O. c. sierra* (p. 79); **i,** Henne's Checkerspot, *O. c. hennei* (p. 79); **j,** Kingston Checkerspot, *O. c. kingstonensis* (p. 79), **k,** Corral Checkerspot, *O. c. corralensis* (p. 79).

PLATE 5. **a,** Crown Fritillary, *Speyeria c. coronis* (p. 75); **b,** Semiramis Fritillary, *S. c. semiramis* (p. 75); **c,** Zerene Fritillary, *S. z. serene* (p. 75); **d,** Malcolm's Fritillary, *S. z. malcolmi* (p. 75); **e,** Colon Checkerspot, *Occidryas colon* (p. 80); **f,** Editha Checkerspot, *O. e. editha* (p. 80); **g,** Quino Checkerspot, *O. e. quino* (p. 81); **h,** Augustina Checkerspot, *O. e. augustina* (p. 81); **i,** Mono Checkerspot, *O. e. monoensis* (p. 80); **j,** Bay Region Checkerspot, *O. e. bayensis* (p. 80); **k,** Cloudborn Checkerspot, *O. e. nubigena* (p. 81); **l,** Monache Checkerspot, *Poladryas arachne monache* (p. 85).

PLATE 6. **a,** Callippe Fritillary, *Speyeria c. callippe* (p. 76); **b,** Nevada Fritillary, *S. c. nevadensis* (p. 76); **c,** Unsilvered Fritillary, *S. a. adiaste* (p. 77); **d,** Atossa Fritillary, *S. a. atossa* (p. 77); **e,** California Patch, *Chlosyne californica* (p. 84); **f,** Gabb's Checkerspot, *Charidryas gabbii* (p. 81); **g,** Acastus Checkerspot, *C. acastus* (p. 82); **h,** Neumoegen's Checkerspot, *C. neumoegeni* (p. 82); **i,** Northern Checkerspot, *C. p. palla* (p. 82); **j,** Malcolm's Checkerspot, *C. damoetas malcolmi* (p. 83); **k,** Hoffmann's Checkerspot, *C. h. hoffmanni* (p. 83); **l,** Crocale Patch, *Chlosyne lacinia crocale* (p. 83).

PLATE 7. a, Tehachapi Fritillary, *Speyeria egleis tehachapina* (p. 77); **b,** Egleis Fritillary, *S. e. egleis* (p. 77); **c,** Irene Fritillary, *S. atlantis irene* (p. 77); **d,** Dodge's Fritillary, *S. a. dodgei* (p. 77); **e,** Arge Fritillary, *S. mormonia arge* (p. 78); **f,** Imperial Checkerspot, *Dymasia chara imperialis* (p. 85); **g,** Leanira Checkerspot, *Thessalia l. leanira* (p. 84); **h,** Cerrita Checkerspot, *T. l. cerrita* (p. 84); **i,** Mountain Crescent, *Phyciodes campestris montanus* (p. 86); **j,** Field Crescent, *P. c. campestris* (p. 86); **k,** Mylitta Crescent, *P. mylitta* (p. 86); **l,** Distinct Crescent, *P. pascoensis distinctus* (p. 85); **m,** Phaon Crescent, *P. phaon* (p. 86).

PLATE 8. **a,** Zephyr Anglewing, *Polygonia zephyrus* (p. 89); **b,** Oreas Anglewing, *P. oreas* (p. 88); **c,** Rustic Anglewing, *P. faunus rusticus* (p. 88); **d,** Satyr Anglewing, *P. satyrus neomarsyas* (p. 87); **e,** California Tortoise Shell, *Nymphalis californica* (p. 90); **f,** Milbert's Tortoise Shell, *Aglais milberti furcillata* (p. 91); **g,** American Painted Lady, *Vanessa virginiensis* (p. 92); **h,** West Coast Lady, *V. annabella* (p. 93); **i,** Painted Lady, *V. cardui* (p. 92); **j,** Dark Buckeye, Dark Peacock, *Junonia nigrosuffusa* (p. 94); **k,** Red Admiral, *Vanessa atalanta rubria* (p. 91); **l,** Buckeye, Peacock Butterfly, *Junonia coenia* (p. 93).

PLATE 9. **a,** Baird's Swallowtail, *Papilio b. bairdii* (p. 98); **b,** Short-tailed Swallowtail, *P. i. indra* (p. 100); **c,** Giant Swallowtail, *P. cresphontes* (p. 101); **d,** Pipe-vine Swallowtail, *Battus p. philenor* (p. 98); **e,** Behr's Parnassian, *Parnassius phoebus behrii* (p. 97); **f,** Edwards's Swallowtail, *Papilio indra pergamus* (p. 100).

PLATE 10. **a,** Pale Swallowtail, *Papilio eurymedon* (p. 102); **b,** Anise Swallowtail, *P. zelicaon* (p. 98); **c,** Western Tiger Swallowtail, *P. rutulus* (p. 101); **d,** Two-tailed Swallowtail, *P. multicaudatus* (p. 102); **e,** Clodius Parnassian, *Parnassius c. clodius* (p. 97); **f,** Wright's Swallowtail, *Papilio polyxenes coloro* (p. 99).

PLATE 11. **a,** Alfalfa Butterfly, *Colias eurytheme* (p. 107); **b,** Clouded Sulfur, *C. philodice eriphile* (p. 107); **c,** Harford's Sulfur, *C. harfordii* (p. 109); **d,** Edwards's Sulfur, *C. alexandra edwardsi* (p. 109); **e,** Behr's Sulfur, *C. behrii* (p. 109); **f,** Cloudless Sulfur, *Phoebis sennae marcellina* (p. 111); **g,** Large Orange Sulfur, *P. argarithe* (p. 111).

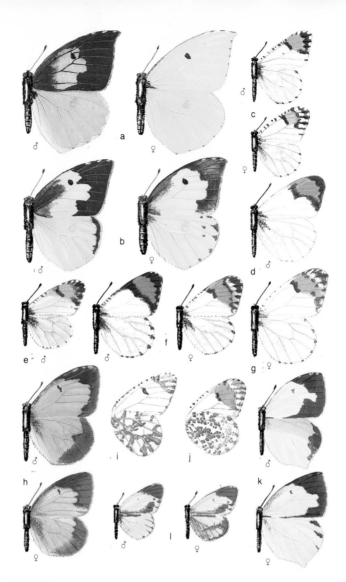

PLATE 12. a, California Dog-face, *Zerene eurydice* (p. 110); **b,** Southern Dog-face, *Z. cesonia* (p. 110); **c,** Felder's Orange-tip, *Anthocharis c. cethura* (p. 113); **d, g,** Sara Orange-tip, *A. s. sara* (p. 113); **e,** Pima Orange-tip, *A. cethura pima* (p. 113); **f, j,** Reakirt's Orange-tip, *A. sara reakirti* (p. 113); **h,** Nicippe Yellow, *Eurema nicippe* (p. 112); **i,** Edwards's Marble, *Euchloe h. hyantis* (p. 115); **k,** Mexican Yellow, *Eurema mexicana* (p. 111); **l,** Dainty Sulfur, *Nathalis iole* (p. 112).

PLATE 13. **a,** Behr's Metalmark, *Apodemia mormo virgulti* (p. 117); **b,** Lange's Metalmark, *A. m. langei* (p. 117); **c,** Palmer's Metalmark, *A. palmeri marginalis* (p. 118); **d,** Wright's Metalmark, *Calephilis wrighti* (p. 119); **e,** Dusky Metalmark, *C. nemesis californica* (p. 118); **f,** Coral Hairstreak, *Harkenclenus titus immaculosus* (p. 121); **g,** Lorquin's Hairstreak, *Habrodais grunus lorquini* (p. 120); **h,** Sarita Hairstreak, *Chlorostrymon simaethis sarita* (p. 120); **i,** Sooty Gossamerwing, *Satyrium fuliginosum* (p. 121); **j,** Behr's Hairstreak, *S. behrii* (p. 121); **k,** Gray Hairstreak, *S. tetra* (p. 122); **l,** Gold-hunter's Hairstreak, *S. a. auretorum* (p. 122); **m,** Nut-brown Hairstreak, *S. a. spadix* (p. 122); **n,** Purplish Hairstreak, *S. saepium chlorophora* (p. 123); **o,** California Hairstreak, *S. californicum* (p. 124); **p,** Woodland Hairstreak, *S. s. sylvinum* (p. 123); **q,** Dryope Hairstreak, *S. dryope* (p. 123); **r,** Leda Hairstreak, *Ministrymon leda* (p. 124); **s,** Ines hairstreak, *M. leda* f. *ines* (p. 125).

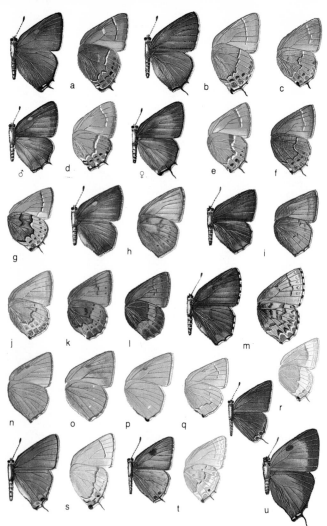

PLATE 14. **a,** Thicket Hairstreak, *Mitoura spinetorum* (p. 127); **b,** Johnson's Hairstreak, *M. johnsoni* (p. 127); **c,** Nelson's Hairstreak, *M. nelsoni* (p. 126); **d,** Siva Hairstreak, *M. s. siva* (p. 125); **e,** Juniper Hairstreak, *M. s. juniperaria* (p. 126); **f,** Muir's Hairstreak, *M. muiri* (p. 127); **g,** Skinner's Hairstreak, *M. loki* (p. 125); **h,** Western Brown Elfin, *Incisalia augusta iroides* (p. 128); **i,** Annette's Brown Elfin, *I. a. annetteae* (p. 129); **j,** Fotis Hairstreak, *I. fotis* (p. 128); **k,** Doudoroff's Hairstreak, *I. mossii doudoroffi* (p. 128); **l,** San Bruno Elfin, *I. m. bayensis* (p. 128); **m,** Western Banded Elfin, *I. eryphon* (p. 129); **n,** Bramble Hairstreak, *Callophrys d. dumetorum* (p. 129); **o.** Green Hairstreak, *C. viridis* (p. 130); **p,** Lembert's Hairstreak, *C. lemberti* (p. 130); **q,** Comstock's Hairstreak, *C. comstocki* (p. 130); **r,** Avalon Hairstreak, *Strymon avalona* (p. 131); **s,** Common Hairstreak, *S. melinus pudicus* (p. 131); **t,** Columella Hairstreak, *S. columella istapa* (p. 132); **u,** Great Purple Hairstreak, *Atlides halesus estesi* (p. 131).

PLATE 15. **a,** Gorgon Copper, *Gaeides gorgon* (p. 134); **b,** Mourning-garbed Copper, *G. xanthoides luctuosa* (p. 134); **c,** Edith's Copper, *G. editha* (p. 134); **d,** Cloudy Copper, *Tharsalea arota nubila* (p. 133); **e,** American Copper, *Lycaena phlaeas hypophlaeas* (p. 133); **f, h,** Ruddy Copper, *Chalceria rubida* (p. 135); **g,** Lustrous Copper, *Lycaena cuprea* (p. 133); **i,** Bright Blue Copper, *Chalceria heteronea clara* (p. 135); **j,** Purplish Copper, *Epidemia helloides* (p. 136); **k,** Varied Blue, *Chalceria h. heteronea* (p. 135); **l,** Nivalis Copper, *Epidemia nivalis* (p. 136); **m,** Mariposa Copper, *E. mariposa* (p. 137); **n,** Hermes Copper, *Hermelycaena hermes* (p. 137).

PLATE 16. a, Marine Blue, *Leptotes marina* (p. 138); **b,** Edwards's Blue, *Hemiargus ceraunus gyas* (p. 139); **c,** Reakirt's Blue, *H. isola alce* (p. 139); **d,** Hilda Blue, *Plebejus saepiolus hilda* (p. 141); **e.** Greenish Blue, *P. s. saepiolus* (p. 141); **f,** Anna Blue, *Lycaeides idas anna* (p. 139); **g, j,** Orange-margined Blue, *L. melissa paradoxa* (p. 140); **h, k,** Pardalis Blue, *Icaricia icarioides pardalis* (p. 142); **i,** San Emigdio Blue, *Plebulina emigdionis* (p. 141); **l,** Evius Blue, *Icaricia icarioides evius* (p. 142); **m,** Shasta Blue, *I. s. shasta* (p. 143); **n,** Acmon Blue, *I. a. acmon* (p. 143); **o,** Clemence's Blue, *I. lupini monticola* (p. 144); **p,** Western Tailed Blue, *Everes amyntula* (p. 145).

PLATE 17. **a,** Gray Blue, *Agriades franklinii podarce* (p. 144); **b,** Square-spotted Blue, *Euphilotes b. battoides* (p. 146); **c,** Dotted Blue, *E. e. enoptes* (p. 146); **d,** Glaucous Blue, *E. b. glaucon* (p. 146); **e,** Dammers's Blue, *E. e. dammersi* (p. 146); **f,** Mojave Blue, *E. mojave* (p. 147); **g,** Elvira's Blue, *E. pallescens elvirae* (p. 147); **h,** Pygmy Blue, *Brephidium exile* (p. 138); **i,** Small Blue, *Philotiella s. speciosa* (p. 148); **j,** Sonoran Blue, *Philotes sonorensis* (p. 148); **k,** Coastal Arrowhead Blue, *Glaucopsyche p. catalina* (p. 150); **l,** Behr's Blue, *G. lygdamus incognitus* (p. 149); **m,** Palos Verdes Blue, *G. l. palosverdesensis* (p. 149); **n,** Xerces Blue, *G. xerces* (p. 150); **o,** Echo Blue, *Celastrina ladon echo* (p. 150); **p,** Cinereous Blue, *C. l. cinerea* (p. 150); **q,** Veined Blue, *Icaricia neurona* (p. 144).

PLATE 18. **a,** Maud's Giant Skipper, *Megathymus coloradensis maudae* (p. 152); **b,** Allie's Giant Skipper, *Agathymus alliae* (p. 151); **c,** Stephens's Giant Skipper, *A. stephensi* (p. 151); **d,** Bauer's Giant Skipper, *A. baueri* (p. 152) (figures **a** to **d** greatly reduced); **e,** Wandering Skipper, *Panoquina errans* (p. 153); **f,** Eufala Skipper, *Lerodea eufala* (p. 155); **g,** Dun Skipper, *Euphyes ruricola* (p. 155); **h,** Umber Skipper, *Paratrytone melane* (p. 156); **i, k,** Verus Farmer, *Ochlodes agricola verus* (p. 157); **j,** Farmer, *O. a. agricola* (p. 157); **l,** Woodland Skipper, *O. sylvanoides* (p. 156); **m,** Yuma Skipper, *O. yuma* (p. 157); **n,** Tecumseh Skipper, *Polites sabuleti tecumseh* (p. 158); **o,** Field Skipper, Sachem, *Atalopedes campestris* (p. 158); **p,** Sandhill Skipper, *Polites s. sabuleti* (p. 158); **q,** Tawny-edged Skipper, *P. themisto-cles* (p. 159); **r,** Chusca Skipper, *P. sabuleti chusca* (p. 158); **s,** Sonora Skipper, *P. sonora sonora* (p. 159); **t,** Dog-star Skipper, *P. s. siris* (p. 159).

PLATE 19. **a,** Harpalus Skipper, *Hesperia comma harpalus* (p. 160); **b,** Yosemite Skipper, *H. c. yosemite* (p. 160); **c,** Oregon Skipper, *H. c. oregonia* (p. 160); **d,** Leussler's Skipper, *H. c. leussleri* (p. 160); **e,** Dodge's Skipper, *H. c. dodgei* (p. 160); **f,** Tilden's Skipper, *H. c. tildeni* (p. 160); **g,** Uncas Skipper, *H. uncas macswaini* (p. 159); **h,** Nevada Skipper, *H. nevada* (p. 161); **i,** Miriam's Skipper, *H. miriamae* (p. 161); **j,** Lindsey's Skipper, H. *lindseyi* (p. 161); **k,** Columbian Skipper, *H. columbia* (p. 162); **l,** Pahaska Skipper, *H. pahaska martini* (p. 162); **m,** Eunus Skipper, *Pseudocopaeodes eunus* (p. 163); **n,** Carus Skipper, *Yvretta carus* (p. 163); **o,** Yuba Skipper, *Hesperia juba* (p. 162); **p,** Hewitson's Skipper, *Copaeodes aurantiaca* (p. 164); **q,** Arctic Skipper, *Carterocephalus palaemon mandan* (p. 166).

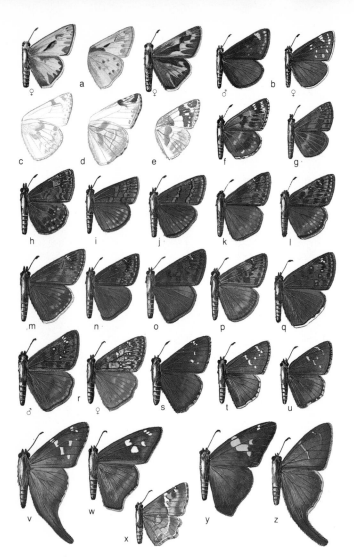

PLATE 20. **a,** Fiery Skipper, *Hylephila phyleus* (p. 163); **b,** Mojave Sooty-wing, *Pholisora libya* (p. 166); **c,** Large White Skipper, *Heliopetes ericetorum* (p. 168); **d,** Laviana Skipper, *H. laviana* (p. 168); **e,** Erichson's Skipper, *H. domicella* (p. 168); **f,** Alpheus Sooty-wing, *Pholisora alpheus oricus* (p. 167); **g,** MacNeill's Sooty-wing, *P. gracielae* (p. 167); **h,** Dreamy Dusky-wing, *Erynnis icelus* (p. 170); **i,** Wright's Dusky-wing, *E. brizo lacustra,* (p. 171); **j,** Burgess's Dusky-wing, *E. b. burgessi* (p. 171); **k,** Persius Dusky-wing, *E. persius* (p. 171); **l,** Afranius Dusky-wing, *E. afranius* (p. 171); **m,** Funereal Dusky-wing, *E. funeralis* (p. 172); **n,** Artful Dusky-wing, *E. pacuvius callidus* (p. 172); **o,** Grinnell's Dusky-wing *E. p. pernigra* (p. 172); **p,** Dyar's Dusky-wing, *E. p. lilius* (p. 172); **q,** Mournful Dusky-wing, *E. tristis* (p. 172); **r,** Propertius Dusky-wing, *E. propertius* (p. 173); **s,** Northern Cloudy-wing, *Thorybes pylades,* (p. 174); **t,** Nevada Cloudy-wing, *T. mexicana nevada* (p. 175); **u,** Diverse Cloudy-wing, *T. diversus* (p. 174); **v,** Long-tailed Skipper, *Urbanus proteus* (p. 175); **w,** Arizona Skipper, *Polygonus leo arizonensis* (p. 176); **x,** Powdered Skipper, *Systasea zampa* (p. 173); **y,** Silver-spotted Skipper, *Epargyreus clarus californicus* (p. 176); **z,** Simplicius Skipper, *Urbanus simplicius* (p. 176).

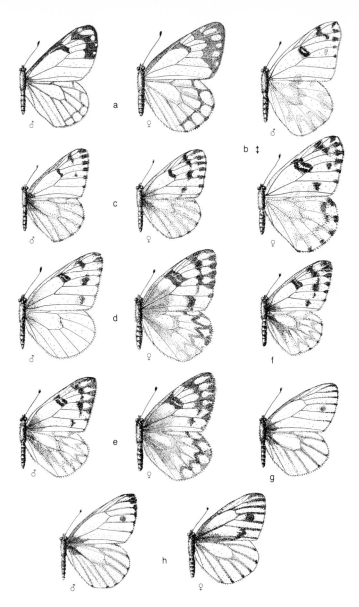

PLATE 21. **a,** Pine White, *Neophasia menapia* (p. 103); **b,** Becker's White, *Pontia beckerii* (p. 104); **c,** California White, *P. sisymbrii* (p. 104); **d,** Common White, *P. protodice* (p. 105); **e,** Western White, *P. o. occidentalis* (p. 105); **f,** Calyce White, *P. o. f. calyce* (p. 106); **g,** Veined White, *Artogeia napi venosa* (p. 106); **h,** Margined White, *A. n. marginalis* (p. 106).

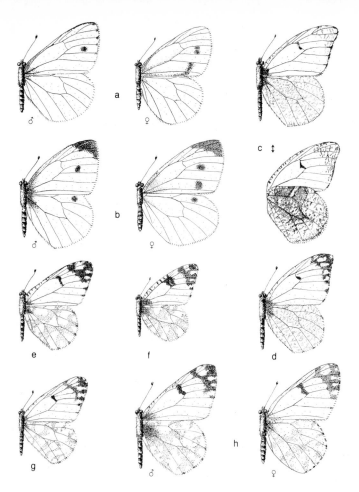

PLATE 22. **a,** Small-veined White, *Artogeia napi microstriata* (p. 106); **b,** Cabbage Butterfly, *A. rapae* (p. 106); **c,** Boisduval's Marble, *Falcapica l. lanceolata* (p. 114); **d,** Grinnell's Marble, *F. l. australis* (p. 114); **e,** Edwards's Marble, *Euchloe h. hyantis* (p. 115); **f,** Southern Marble, *E. h. lotta* (p. 115); **g,** Martin's marble, *E. h. andrewsi* (p. 115); **h,** Large Marble, *E. ausonides* (p. 115).

PLATE 23. **a,** Behr's Parnassian, *Parnassius phoebus behrii* (p. 97); **b,** Clodius Parnassian, *P. c. clodius* (p. 97); **c,** Mourning Cloak, *Nymphalis antiopa* (p. 90); **d,** Wide-banded Admiral, *Basilarchia weidemeyerii latifascia* (p. 94); **e,** California Sister, *Adelpha bredowii californica* (p. 95); **f,** Lorquin's Admiral, *Basilarchia lorquini* (p. 95); **g,** Mormon Metalmark, *Apodemia m. mormo* (p. 117); **h,** Lange's Metalmark, *A. m. langei* (p. 117); **i,** Behr's Metalmark, *A. m. virgulti* (p. 117); **j,** Desert Metalmark, *A. m. deserti* (p. 117); **k,** Cythera Metalmark, *A. m. cythera* (p. 117); **l,** Whitish Metalmark, *A. m. dialeuca* (p. 118); **m,** Palmer's Metalmark, *A. palmeri marginalis* (p. 118).

PLATE 24. **a,** Allie's Giant Skipper, *Agathymus alliae* (p. 151); **b,** Bauer's Giant Skipper, *A. baueri* (p. 152); **c,** Wandering Skipper, *Panoquina errans* (p. 153); **d,** Eufala Skipper, *Lerodea eufala* (p. 155); **e,** Roadside Skipper, *Amblyscirtes vialis* (p. 155); **f,** Ceos Sooty-wing, *Staphylus ceos* (p. 174); **g,** Julia's Skipper, *Nastra julia* (p. 164); **h,** Common Sooty-wing, *Pholisora catullus* (p. 166); **i,** Mojave Sooty-wing, *P. l. libya* (p. 166); **j,** Large White Skipper, *Heliopetes ericetorum* (p. 168); **k,** Laviana Skipper, *H. laviana* (p. 168); **l,** Erichson's Skipper, *H. domicella* (p. 168); **m,** Little Checkered Skipper, *Pyrgus scriptura* (p. 169); **n,** Rural Skipper, *P. r. ruralis* (p. 169); **o,** Common Checkered Skipper, *P. communis* (p. 169); **p,** Western Checkered Skipper, *P. albescens* (p. 170).

4 · CALIFORNIA BUTTERFLIES: SPECIES ACCOUNTS

Abbreviations

Cu_2	cubitus 2 vein
el.	elevation
f.	form
ft.	feet
FW	forewing
HW	hind wing
Hwy.	highway
in.	inches
mi.	miles
mm	millimeters
Mt.	Mount
nr.	near
sp., spp.	species (sing., pl.)
TL	type locality
UN	underside
UP	upper side
U.S.	United States
var.	variety

Satyrs, Arctics, and Ringlets
(Family Satyridae)

Small to medium-sized, 1¼–2¾ in. (31–69 mm), dull-colored butterflies, often with eyespots (round, eyelike markings) on the wings, and with some of the veins of FW swollen. *Egg* dome shaped, flattened on top. *Larva* cylindrical, smooth, with the posterior end notched. *Pupa* smooth, suspended head downward by posterior end. Flight of adults bounding.

California Ringlet   (*Coenonympha california*). This plain-colored ringlet is one of our commonest butterflies, and one of the least conspicuous. The typical subspecies (*C. c. california*) (Pl. 3i) is found almost everywhere in California west of the Sierra Nevada and the Mojave and Colorado deserts, in grassland or oak woodland, and may be recognized by its low, bouncing flight. The Siskiyou Ringlet (*C. c. eryngii*), found from Tehama County northward into Oregon, is lighter UN. Flight period February–October in lowland and median elevations; at higher elevations, emergence is delayed and brood numbers decrease from three to two. *Early stages: egg* dome shaped, flattened on top, with a small button in the center; *larva* pale green streaked with brown, posterior end notched; *pupa* pale green or brown. *Larval food plants:* grasses (Poaceae).

Ringless Ringlet   (*Coenonympha ampelos elko*) (Pl. 3h). Pale buffy, outer third UNHW very light, all eyespots lacking or very small. Flight behavior similar to that of California Ringlet. Found in northern California and Nevada, east of the Cascades, and northward. *Early stages:* very similar to those of the California Ringlet. *Larval food plants:* grasses (Poaceae).

Ochraceous Ringlet   (*Coenonympha ochracea*) (Pl. 3j). A big yellowish tan ringlet, found from the Rocky Mountains west into California east of the Sierra Nevada; two broods. The Californian subspecies, the Mono Ringlet (*C. o. mono*), flies in the wet meadows around Bridgeport and Mono Lake from late June to early September. *Early stages:* unrecorded. *Larval food plants:* undoubtedly grasses (Poaceae).

Riding's Satyr   (*Neominois ridingsii*) (Pl. 2i). This strange "waif of the mountains," gray, with an irregular light band across the wings, and two black spots, flies with the Ivallda Arctic on the higher Sierra Nevada peaks from Inyo County north to Carson Pass and beyond, and also in the White Mountains of Inyo and Mono counties. It flushes from underfoot, like a grasshopper or a moth, only to disappear from sight upon alighting on a lichen-covered rock. Flight period July–August; one brood. *Early stages: egg* white, barrel shaped, ribbed; *larva* reddish buff, sides pale green with a lengthwise black stripe; *pupa* reddish brown, wing cases green; pupation underground. *Larval food plants:* grasses (Poaceae).

Wood Nymph   (*Cercyonis pegala*). This dark, unusual-looking butterfly is found in some subspecies from the Atlantic to the Pacific and from Canada to Florida and Texas. The lower FW eyespot is larger than the upper one. California has two subspecies, each with two or more forms: (1) Ox-eyed Satyr (*C. p. boopis*) (Pl. 2d), eyespots UNHW small to absent, found in most of northern California; Baron's Satyr (form "*baroni*") (Pl. 2e), UNHW eyespots usually well developed, found in northern coastal region; Hoary Satyr (form "*incana*") (Pl. 2f), UNHW often frosted with whitish overlay, found in inner ranges from Modoc and Siskiyou counties northward. (2) Ariane Satyr (*C. p. ariane*), eyespots of UNHW in two groups of three, the center spot of each group largest, all spots usually lightringed, females often light colored, found in low, often alkaline, Great Basin flats from eastern California to Utah; Stephens's Satyr (form "*stephensi*") (Pl. 2g), very light colored, females particularly so, but males also lighter than normal *ariane*, found in alkaline flats of eastern California and in Nevada, regarded as a subspecies by some. Flight period late June–August; one brood. *Early stages: egg* barrel shaped, ribbed; *larva* pale green with four light stripes; *pupa* pale green. *Larval food plants:* grasses (Poaceae).

Sthenele Satyr   (*Cercyonis sthenele*). Much smaller than the Wood Nymph. Upper FW eyespot larger than lower; both eyespots about the same distance from the wing edge. Three

California subspecies: (1) Sthenele Satyr (*C. s. sthenele*), dark band UNHW; formerly found on the San Francisco Peninsula, now extinct. (2) Woodland Satyr, Sylvan Satyr (*C. s. silvestris*) (Pl. 2a), UNHW without dark band, and often without eyespots; found west of the Sierran crest in northern California and in the Transverse and Coast ranges of southern California, usually in oak woodland. (3) Little Satyr (*C. s. paula*) (Pl. 2b), usually with well-developed eyespots UNHW; northern California east of the Sierra Nevada, east to the Great Basin of Utah, southern California throughout the eastern Mojave Desert ranges, usually in pinyon-juniper woodland. Flight period May–June at low elevations, late June–August northerly and east of the Sierra Nevada; one brood. *Early stages: egg* white, barrel shaped; *larva* green with dark dorsal and yellow lateral stripe; *pupa* green. *Larval food plants:* grasses (Poaceae).

Least Satyr (*Cercyonis oeta*) (Pl. 2c). A neat little satyr, UN with dark dashes, eyespots lightringed; FW narrow, lower FW eyespot smaller and nearer wing margin than upper. Found from Arizona north into British Columbia; in California, east of the Sierra Nevada from Inyo County north. Characteristic of the sagebrush (*Artemisia*) community. Flight period late June–August; one brood. *Early stages: egg* barrel shaped, green, ribbed; *larva* green, tapering to both ends, tails tipped with red; *pupa* light green to black, often conspicuously striped. *Larval food plants:* grasses (Poaceae).

Great Arctic (*Oeneis nevadensis*). A large, bright orange brown arctic, with light areas nearly unmarked. It flies in openings in coniferous forest. California has two subspecies: (1) Great Arctic (*O. n. nevadensis*) (Pl. 2l); Sierra Nevada from Plumas County north. (2) Iduna Arctic (*O. n. iduna*) (Pl. 2k), UP lighter, primaries more pointed, discal band UNHW less scalloped; Coast Ranges from Sonoma County north to southern Oregon. Flight period late May–July; one brood. The Great Arctic has a two-year cycle, is more common in even-numbered years. *Early stages: egg* grayish white, barrel shaped, ribbed; *larva* brownish buff with a black middorsal stripe; *pupa* undescribed. *Larval food plants:* grasses (Poaceae).

Ivallda Arctic   (*Oeneis .ivallda*) (Pl. 2h). This medium-sized, dull yellow brown arctic looks undistinguished at a distance, but close up it is very neatly marked. It is found in the Sierra Nevada, from Inyo County north to Donner Pass, disporting around cliffs and summits, where it blends with the granite rocks that surround it. Like other arctics, it leans almost flat upon alighting, perhaps for camouflage, or to make use of available heat. Flight period late June–August; one brood. *Early stages: egg* and young *larva* have been illustrated; mature larva and *pupa* unknown. *Larval food plants:* grasses (Poaceae).

Chryxus Arctic   (*Oeneis chryxus stanislaus*) (Pl. 2j). FW dark at base, buffy on outer third; UNHW strongly marked light and dark. All of our arctics are collector's prizes, but the California subspecies of the Chryxus Arctic is particularly so, as it is limited to the general vicinity of Sonora Pass in the Sierra Nevada. This is an area of brown volcanic rock, with which the Chryxus Arctic blends. Other subspecies are found from Michigan west to New Mexico and north to southern Alaska. The Chryxus Arctic divides the range of the Ivallda Arctic, which is found both north and south of it. Flight period July–August; one brood. *Early stages*: undescribed for the California subspecies; *larva*, as described for *O. c. chryxus*, greenish brown, striped, short-hairy. *Larval food plants*: grasses (Poaceae).

Milkweed Butterflies (Family Danaidae)

Large, brownish butterflies, 2⅝–4 in. (66–100 mm), conspicuously marked with black and white. Front legs reduced, held against the thorax, a character shared with Satyridae, Nymphalidae, and Libytheidae. Found in both Old and New World, mostly tropical. Larvae feed on milkweed, obtaining substances distasteful to birds and mammals. Adults are mimicked by certain members of nonprotected families. Two species occur in California.

Monarch   (*Danaus plexippus*) (Pls. 1k, 1l, 3a). One of our most conspicuous and best-known butterflies, the Monarch belongs to a tropical family and cannot live through northern winters. Adults migrate southward in the fall and pass the winter at many places along the California coast, including Pacific Grove on the Monterey Peninsula, where they are expressly protected. Here thousands may be seen on the famous butterfly trees, an important tourist attraction. In overwintering areas, they fly about only during periods of warm, sunny weather. When spring returns, they move northward, some of their descendents eventually reaching southern Canada. The individuals that return in the fall are descendents of those that migrated northward in the spring. Flight is powerful, lofty, and soaring. There are several broods a year. *Early stages*: *egg* pale green, higher than wide, with vertical ribs; *larva* dull green or whitish, ringed with yellow and black, fleshy horns (filaments) at each end; *pupa* short and stubby, pale green with golden spots and studlike projections, suspended by the posterior end. *Larval food plants*: true milkweeds (*Asclepias* spp.) (Asclepiadaceae), including Narrow-leaved Milkweed (*A. fascicularis*) and Indian Milkweed (*A. eriocarpa*) in cismontane and southern California.

Striated Queen   (*Danaus gilippus strigosus*) (Pl. 3b). Dark reddish brown, much darker than the orange brown Monarch. Common in Arizona, east to Texas, and in the Colorado Desert of eastern Riverside and Imperial counties. It occurs as a straggler east of the Sierra Nevada in Inyo and Mono counties. Occasional in the fall coastally from Santa Barbara to San Diego counties. Flight period April–November; a succession of broods. *Early stages*: *egg* pale green; *larva* dull white with rings of purple, reddish brown, and yellow, has three pairs of horns (filaments) (one more pair than the Monarch); *pupa* green with golden spots, suspended by posterior end. *Larval food plants*: climbing milkweed (*Sarcostemma* spp., including *S. hirtellum* and *S. cynanchoides* var. *hartwegii* in California), also true milkweeds (*Asclepias* spp.), including White-stemmed Milkweed (*A. albicans*) and Desert Milkweed (*A. erosa*) (Asclepiadaceae).

Long-wings (Family Heliconiidae)

Forewing long, narrow. Antennae long, body slender. Otherwise as in Nymphalidae, of which they are by some considered a subfamily. Neotropical, north to southern United States. One species in California.

Gulf Fritillary   (*Agraulis vanillae incarnata*) (Pl. 3f). 2–3 in. (50–75 mm). This striking species, known by its long forewings, the orange brown color, and the brilliant silver spots UNHW, is our only representative of the Heliconiidae, a tropical family. Formerly scarce north of Santa Barbara, it has become locally common in central California, and has been found as far north as Redding, Shasta County. Flight period March–November; several broods a year. *Early stages*: *egg* pale green, larger at top than bottom, like an inverted flask; *larva* slender, purple-and-orange striped, with erect, widely spaced spines; *pupa* dark brown, oddly shaped, with very large, bulging wing covers. *Larval food plants*: introduced and cultivated species of passion vine (*Passiflora* spp.) (Passifloraceae).

Brush-footed Butterflies (Family Nymphalidae)

Small to large butterflies, 1⅛–3⅜ in. (28–84 mm). Front legs greatly reduced, held appressed to the body. Discal cell of hind wing open; antennae scaled above. *Egg* rounded or dome shaped, with vertical ribs. *Larva* spiny, including head. *Pupa* suspended from posterior end, often rough or knobby. A very large family, including many of our best-known butterflies, some of powerful flight.

Variegated Fritillary   (*Euptoieta claudia*) (Pl. 3e). Like the Gulf Fritillary, the Variegated Fritillary invades our region from the Gulf States and Mexico. Smaller in size, it lacks the silver spots UNHW, and the elongate forewing of that species. Rarely encountered in southern California, it flies in the New York and Providence mountains of eastern San Bernardino County, and at Scissors Crossing, San Diego County. Flight period June–September; multiple brooded. *Early stages*: *egg* conical, pale green; *larva* reddish yellow with black spines and brown bands; *pupa* white with black blotches. *Larval food plants*: unknown for California, but believed to be flax (*Linum* spp.) (Linaceae); elsewhere, passion vine (*Passiflora* spp.) (Passifloraceae); may become a pest on pansies (*Viola* spp.) (Violaceae).

Fritillaries of the genus *Speyeria* have certain characteristics in common. All are single brooded, and females tend to emerge later than males. Their flight periods differ, some species, like the Callippe Fritillary (*S. callippe*), flying in the spring and early summer and others, like the Nokomis Fritillary (*S. nokomis*), flying in the late summer and early fall. The early stages of all are similar, and the few Californian species for which these are unknown or as yet undescribed are probably not very different from the others. The larvae, which are nocturnal, feed on violets (*Viola* spp.) (Violaceae).

Leto Fritillary   (*Speyeria cybele leto*) (Pl. 4c). A large, beautiful silverspot (another name for fritillary), the male bright orange brown, the female black with a light straw-colored submarginal band. UNHW, wing bases dark, submarginal band light; silver spots small and well separated. Leto flies at middle elevations both east and west of the Sierra-Cascade crest, often in aspen groves. Flight period July–August; one brood. *Early stages*: *larva* dark and spiny. *Larval food plants*: various species of violets (*Viola* spp.) (Violaceae).

Apache Fritillary   (*Speyeria nokomis apacheana*) (Pl. 4b). Our largest silverspot, even larger and brighter than Leto. The male is a bright reddish-tawny, the female dark like that of Leto. UNFW has a basal reddish flush; UNHW is yellowish in male, greenish with light submarginal band in female. Silver spots large, dark-edged. This regal species is found in wet meadows, and can seldom be taken without some wading. Found east of the Sierra Nevada, mostly in Inyo, Mono, and Alpine counties, and in Nevada. Flight period late, in late August and September; males precede females by two weeks; one brood. *Early stages*: *egg* cream color to tan, ribbed; mature *larva* orange ochre with black spots and lines, spines brown with black tips; *pupa* orange ochre with black markings, wing cases lighter. *Larval food plants*: (*Viola* spp.) (Violaceae).

Crown Fritillary   (*Speyeria coronis*). This spangled beauty was first found in the Coast Ranges. The TL, Alma, Santa Clara County, is now deep under the waters of Lexington Reservoir. Compared to Zerene and Callippe, it is lighter, both UP and UN; spots large, well-silvered. Found in canyons and woodland clearings, and, east of the Sierra Nevada, in sagebrush and forest edges. Flight period for males, June–July, females often as late as September. California has five recognizably similar subspecies: (1) Crown Fritillary (*S. c. coronis*) (Pl. 5a), Coast Ranges from San Luis Obispo County north. (2) Henne's Fritillary (*S. c. hennei*), Tejon and Tehachapi Mountains, Ventura and Kern counties. (3) Semiramis Fritillary (*S. c. semiramis*) (Pl. 5b), redder, more lightly marked, San Gabriel and San Bernardino mountains, south to Mexico. (4) Simaetha Fritillary (*S. c.* nr. *simaetha*) (TL Black Canyon nr. Brewster, Brewster County, Washington), Trinity Alps. (5) Snyder's Fritillary (*S. c. snyderi*), east of Sierra Nevada, east to Utah.

Zerene Fritillary   (*Speyeria zerene*). Usually bright orange brown; dark markings heavy. Found in main forest belt, in both Sierra Nevada and Coast Ranges. Three of California's seven subspecies were formerly considered distinct species: (1) Zerene Fritillary (*S. z. zerene*) (Pl. 5c), UNHW purplish brown, spots unsilvered; west of Sierra Nevada crest, and inner Coast Ranges. (2) Royal Fritillary (*S. z. conchyliatus*), similar, but UN very dark; Shasta and Siskiyou counties. (3) Malcolm's Fritillary (*S. z. malcolmi*) (Pl. 5d), much lighter; east slope of Sierra Nevada, centering around Mono County. (4) Behrens's Fritillary (*S. z. behrensii*), similar, but UNHW rich dark brown;

Mendocino and Humboldt counties. (5) Myrtle's Fritillary (*S. z. myrtleae*), UN golden brown, spots small, brightly silvered; coastal Marin and San Mateo counties. (6) Gunder's Fritillary (*S. z. gunderi*) (= *cynna* of lists), very light colored, almost golden; UNHW spots brightly silvered; Warner Mountains of Modoc County, Ruby Mountains of Nevada, Steens Mountains of Oregon. (7) Glorious Fritillary (*S. z. gloriosa*), similar, but somewhat lighter; Del Norte County, and Josephine County, Oregon.

Callippe Fritillary (*Speyeria callippe*). Our most evident fritillary, found on open hillsides where Wild Pansy (*Viola pedunculata*) grows. UP black markings narrow; UNHW marginal silver spots pyramid-shaped, tips pointed inward. California has no less than nine rather different-looking subspecies: (1) Callippe Fritillary (*S. c. callippe*) (Pl. 6a), dull yellowish brown with sooty markings; now scarce or extinct in much of its former range, except San Bruno Mountains. (2) Lilian's Fritillary (*S. c. liliana*), UP bright orange brown, UNHW submarginal band yellowish; Coast Ranges north of San Francisco Bay. (3) Rupestris Fritillary (*S. c. rupestris*), UP dull orange brown, UNHW spots usually unsilvered; Trinity County. (4) Yuba Fritillary (*S. c. juba*), UNHW smooth purplish brown, spots often only partly silvered; west slope Sierra Nevada, north to Tehama County. (5) Sierra Fritillary (*S. c. sierra*), small, UNHW bright buffy, spots brightly silvered; Sierra Nevada in Sierra and Plumas counties; grades gradually into Nevada Fritillary on east slope. (6) Nevada Fritillary (*S. c. nevadensis*) (Pl. 6b), UNHW tinted green; east of Sierra-Cascade crest; Nevada. (7) Comstock's Fritillary (*S. c. comstocki*), light colored, UNHW tawny; Coast Ranges from Santa Clara County southward, including Santa Cruz Island. (8) Macaria Fritillary (*S. c. macaria*), light colored, UNHW buffy; interior ranges of southern California, from Alamo Mountain, Ventura County, north to southern Sierra Nevada above Kernville; unsilvered form known as Laurina Fritillary (form "*laurina*") or Unsilvered Macaria. (9) Plain Fritillary (*S. c. inornata*), UNHW dull brown with a slight glaucous sheen; spots dull yellowish, unsilvered; Sierra Nevada foothills from Fresno County north to blend with *S. c. juba*.

Unsilvered Fritillary   (*Speyeria adiaste*). This pretty species is found in meadows and clearings. It is smaller and more delicate than *Callippe*, with reduced markings and unsilvered UNHW spots. Flight season June–July. There are three subspecies: (1) Unsilvered Fritillary (*S. a. adiaste*) (Pl. 6c), bright brownish red; closely associated with redwood forests south of San Francisco Bay, in San Mateo and Santa Cruz counties. (2) Clemence's Fritillary (*S. a. clemencei*), intermediate in color and markings; Coast Ranges of Monterey and San Luis Obispo counties. (3) Atossa Fritillary (*S. a. atossa*) (Pl. 6d), larger, paler, UNHW creamy buff; Tejon and Tehachapi ranges of Ventura, Kern, and Los Angeles counties, now probably extinct.

Egleis Fritillary   (*Speyeria egleis*). Formerly known as the Mountain Vagabond (*S. montivaga*), this little species is characteristic of the slopes and meadows of the Sierra Nevada and Cascades. *Egleis* flies in midsummer and is usually common where it occurs. It is quite variable. California has three subspecies: (1) Egleis Fritillary (*S. e. egleis*) (Pl. 7b), UP tawny, the dark markings slight, UNHW dull buff mixed with brown; Sierra Nevada at high elevations. (2) Owen's Fritillary (*S. e. oweni*), UP markings heavy, UNHW brown; Mt. Shasta and vicinity, north into Oregon. (3) Tehachapi Fritillary (*S. e. tehachapina*) (Pl. 7a), small; UNHW dull, often unsilvered; found on a few of the highest peaks of the Tehachapi and Piute mountains, Kern County.

Atlantis Fritillary   (*Speyeria atlantis*). Though considered a subspecies of the Atlantis Fritillary, our small, lightly marked Irene Fritillary (*S. a. irene*) (Pl. 7c) is very different in appearance; in fact, it looks more like a smaller, paler understudy of the Hydaspe Fritillary. It is found in the central Sierra Nevada north to Lassen County at intermediate elevations. In northern California a second subspecies, Dodge's Fritillary (*S. a.*

dodgei) (Pl. 7d) is found. It is larger, darker, and more heavily marked than Irene, but equally unsilvered; found in Trinity Alps, and Siskiyou Mountains north into Oregon (TL Crater Lake National Park).

Hydaspe Fritillary   (*Speyeria hydaspe*). A conspicuous element of the butterfly fauna of the Sierra Nevada and Cascade Mountains, Hydaspe is a bright orange brown butterfly with heavy markings. Three subspecies occur in California: (1) Hydaspe Fritillary (*S. h. hydaspe*) (Pl. 4d), UNHW light brown, the spots large, buffy, and unsilvered; west slope of the Sierra Nevada at middle elevations. (2) Purple Fritillary (*S. h. purpurascens*), UNHW deep purplish brown, the spots ivory; Lassen and Siskiyou counties. (3) Greenhorn Fritillary (*S. h. viridicornis*), more lightly colored but distinctly marked, UNHW increasingly buffy; Greenhorn Mountains of northern Kern County. Fresh specimens of Hydaspe are very handsome. Flight period: late June–August.

Mormon Fritillary   (*Speyeria mormonia*). Slightly smaller than the Egleis Fritillary, with which it often flies, and with the dark marginal markings slightly heavier, the Mormon Fritillary can be distinguished by the narrow front wing veins of the male, those of the Egleis Fritillary being wider because of sex scaling. California has two subspecies: (1) Arge Fritillary (*S. m. arge*) (Pl. 7e), very small and neat; high elevations in the Sierra Nevada. (2) Erinna Fritillary (*S. m. erinna*), slightly larger, darker; Lassen County north into Oregon. The Mormon Fritillary flies in July–August.

Western Meadow Fritillary   (*Clossiana epithore*) (formerly *Boloria*). Resembles a small silverspot. The Meadow Fritillary is a bright orange brown, with dark markings small and separate. UNHW purplish brown, the spots unsilvered. California has three subspecies: (1) Western Meadow Fritillary (*C. e. epithore*) (Pl. 4a), color as above; Santa Cruz and San Mateo counties. (2) Chermock's Meadow Fritillary (*C. e. chermocki*), smaller, darker, especially UP wing bases; northern Sierra Nevada, Coast Ranges north of San Francisco Bay, and northward. (3) Sierra Meadow Fritillary (*C. e. sierra*), smallest; bright orange brown, markings small and crowded; central and southern Sierra Nevada at middle elevations. Flight period mid-May (low elevations) to July (higher elevations); one brood. *Early stages*: apparently unknown; *larva* supposedly feeds on violets (*Viola* spp.) (Violaceae).

Common Checkerspot, Chalcedon Checkerspot (*Occidryas chalcedona*) (formerly *Euphydryas*). In much of California this common but elegant species may be recognized by its black-and-white checkering, its narrow red trim, and its pointed forewings. However, in the central Sierra Nevada and south to eastern San Bernardino County, there are subspecies that are almost brick red. The Common Checkerspot is usually quite tame, and at times may be picked from flowers by hand. It is extremely variable, and sometimes aberrations occur that have more black or white than normal. Flight period as early as March (low elevations south and east) to June–July (higher elevations); one brood. *Early stages*: *egg* domeshaped, ribbed, yellow at first, later turning red; *larva* black, spiny, colonial on food plants; *pupa* ivory to lavender gray, spotted with black and orange. Eight subspecies are found in California: (1) Common Checkerspot (*O. c. chalcedona*) (Pl. 4e), Coast Ranges and Sierra Nevada foothills. (2) Dwinelle's Checkerspot (*O. c. dwinellei*) (Pl. 4f), considerable red UPFW; local near Bartle, south Siskiyou County. (3) McGlashan's Checkerspot (*O. c. macglashanii*), white checkering larger; northern Sierra Nevada. (4) Sierra Checkerspot (*O. c. sierra*) (Pl. 4h), south and central Sierra Nevada at high elevations. (5) Olancha Checkerspot (*O. c. olancha*) (Pl. 4g), white-and-red checkering about equal to black areas; east slope of southern Sierra Nevada. (6) Henne's Checkerspot (*O. c. hennei*) (formerly listed as *O. c. quino*, a name correctly applied to a subspecies of *Occidryas editha*) (Pl. 4i), small, less red color, white markings prominent; desert canyons and washes of Riverside and San Diego counties. (7) Kingston Checkerspot (*O. c. kingstonensis*) (Pl. 4j), UP salmon-and-black checkered, UN tawny, white-spotted; Kingston Mountains and adjacent ranges of San Bernardino County, south to Granite Mountains. (8) Corral

Checkerspot (*O. c. corralensis*) (Pl. 4k), quite red, dark markings narrow; east end of San Bernardino Mountains. *Larval food plants*: members of the Figwort Family (Scrophulariaceae), including California Bee Plant (*Scrophularia californica*), monkey flowers (*Mimulus* spp.), paintbrushes (*Castilleja* spp.), and others.

Colon Checkerspot   (*Occidryas colon*) (formerly *Euphydryas*) (Pl. 5e). Very much like the Common Checkerspot, of which it has at times been considered a subspecies. UP blacker because of small size of white markings; UN brick red, the white bands narrow and regular, the black edgings narrow. Northern Siskiyou and Modoc counties (Warner Mountains), north to British Columbia, east to Idaho. Flight period June–July; one brood. *Early stages*: *eggs* laid in clusters; *larva* covered with fuzzy white hairs; *pupa* unknown. *Larval food plant*: Common Snowberry (*Symphoricarpos rivularis*) (Caprifoliaceae).

Editha Checkerspot   (*Occidryas editha*) (formerly *Euphydryas*). More local and colonial than the Common Checkerspot. Usually found on metamorphic soils such as serpentine, and flies early in the year at low elevations; one brood. Editha has short, rounded forewings, and strong red bands UP. It is split into numerous subspecies, some of which were formerly considered full species and have long had common names. California has no less than twelve subspecies: (1) Editha Checkerspot (*O. e. editha*) (Pl. 5f), red bands nearly brick red, light bands distinct; lower mountains and foothills of Kern, Tulare, and Fresno counties, on west slope of Sierra Nevada; meets *O. e. rubicunda* in Madera County. (2) Bay Region Checkerspot (*O. e. bayensis*) (Pl. 5j), from Twin Peaks, San Francisco, south to northern Santa Clara County; now greatly reduced in numbers, due to settlement of the area. (3) Luesther's Checkerspot (*O. e. luestherae*), red markings UP wider and brighter; more red in basal bands UNHW; Mt. Hamilton Range south to San Luis Obispo County. (A disjunct population, formerly referred to *bayensis*, but here left unnamed, occurs in the Santa Monica Mountains). (4) Mono Checkerspot (*O. e. monoensis*) (Pl. 5i), red bands rusty, wide; east side of Sierra Nevada, centering in Mono County, now very scarce. (5) Ruddy Checkerspot (*O. e. rubicunda*), red bands very bright; west slope of Sierra Nevada foothills. (6) Cloud-born Checkerspot (*O. e. nubigena*) (Pl. 5k), very small (1.25–1.4 in.), quite red, the dark markings narrow; very high elevations in central Sierra Nevada. (7) Gold Lake Checkerspot (*O. e. aurilacus*), mostly red; dark markings heavy only near wing bases; central mass of Sierra Nevada and southern Cascades. (8) Baron's Checkerspot (*O. e. baroni*), Coast Ranges north of San Francisco Bay. (9) Strand's Checkerspot (*O. e. edithana*), little known; Lassen County north into at least Klamath County, Oregon. (10) Augustina Checkerspot (*O. e. augustina*) (formerly *augusta*, a synonym of *quino*) (Pl. 5h), red-and-yellow bands expanded, size small; San Bernardino Mountains at higher elevations. (11) Quino Checkerspot (*O. e. quino*) (formerly *wrighti*, a synonym of *quino*) (Pl. 5g), large, reddish; coastal ranges of San Diego, Orange, and Riverside counties. (12) Island Checkerspot (*O. e. insularis*), a Santa Rosa Island segregate of the disjunct Santa Monica Mountains population, referred to above. *Early stages*: similar to those of the Common Checkerspot. *Larval food plants*: Dwarf Plantain (*Plantago erecta*) (Plantaginaceae), Owl's Clover (*Orthocarpus densiflorus*), and Indian Warrior (*Pedicularis densiflora*) (both Scrophulariaceae) in the Coast Ranges; in the Sierra Nevada and San Bernardino Mountains, also Chinese Houses (*Collinsia heterophylla*) (Scrophulariaceae); no doubt others in these families.

Gabb's Checkerspot (*Charidryas gabbii*) (formerly *Chlosyne*) (Pl. 6f). Very similar to the Northern Checkerspot; dark markings less heavy; UNHW white bands with a distinctive pearly luster. Southern California north to Monterey County. Flight period March–July; one brood. *Early stages*: *egg* subglobular, pale green with yellow tinge, laid in masses; young *larvae* gregarious; mature larvae black with short branching spines and scattered orange-and-white spots; *pupa* ivory with brown-and-black blotches, a series of orange tubercles dorsally. *Larval food plants*: Beach Aster (*Corethrogyne filaginifolia*) (Asteraceae), perhaps also other Asteraceae.

Neumoegen's Checkerspot   (*Charidryas neumoegeni*) (formerly Chlosyne) (Pl. 6h). This bright orange red checkerspot with the pearly white underside flies on rocky buttes and in canyons and washes in the Mojave and Colorado deserts. It occurs from northern Los Angeles and southern Kern counties through the eastern Mojave Desert ranges into Inyo, San Bernardino, and Riverside counties to Nevada and Arizona. Less common in San Diego and Imperial counties, south into Baja California. Flight period early spring (March–April), one brood; with summer rainfall a second brood in fall. *Early stages*: *egg* pale green; *larva* black, spiny; *pupa* black, marked with gray. *Larval food plant*: Desert Aster (*Machaeranthera tortifolia*) (Asteraceae), on which it can easily be raised.

Acastus Checkerspot   (*Charidryas acastus*) (formerly *Chlosyne*) (Pl. 6g). A Great Basin species that enters California east of the Sierra Nevada. Much like the Northern Checkerspot, but larger, lighter, more tawny. Flies in May and June in our area; one brood in California, 2–3 broods elsewhere. *Early stages*: similar to those of the Northern Checkerspot. *Larval food plant*: Sticky-leaved Rabbit Brush (*Chrysothamnus viscidiflorus*) (Asteraceae).

Northern Checkerspot   (*Charidryas palla*) (formerly *Chlosyne*). Rusty red, the wing bases much darker. Darker than Gabb's Checkerspot; the white bands UNHW lacking pearly luster. The female may be like the male, or very much darker, sometimes black with scattered white spots. Found in canyons, meadows, forest openings. Flight period May–June (July at high elevations); one brood. Subspecies: (1) Northern Checkerspot (*C. p. palla*) (Pl. 6i), found in much of northern California at low to moderate elevations. (2) Whitney's Checkerspot (*C. p. whitneyi*), much redder and less heavily marked; high elevations in the central Sierra Nevada. (3) Death Valley Checkerspot (*C. p. vallismortis*), larger; lighter orange brown, with heavier dark bands UP; Panamint Range, Death Valley National Monument. *Early stages*: only partly known. *Larval food plants*: of *C. p. palla*, long stated to be paintbrush (*Castilleja* spp.) (Scrophulariaceae), but larvae found by J. Emmel,

Shields, and Breedlove on California Aster (*Aster californi-cus*); of *C. p. whitneyi*, rabbit brush (*Chrysothamnus* spp.) (both Asteraceae).

Malcolm's Checkerspot (*Charidryas damoetas mal-colmi*) (formerly *Chlosyne*) (Pl. 6j). Light orange brown, the black markings narrow, but wing bases heavily clouded; in slanted light, a slight sheen is detectable. Small, short-winged. Flight period July–August; one brood. Found in alpine fell-fields of highest Sierran peaks. *Early stages*: unknown. *Larval food plant*: unknown in California; aster (Asteraceae) suspected.

Hoffmann's Checkerspot (*Charidryas hoffmanni*) (for-merly *Chlosyne*). Appearance distinctive, the dark wing bases separated from the rusty wing tips by a pale median band. A local, usually uncommon, butterfly, found in meadows and openings in coniferous forests. Flight period July–early Au-gust; one brood. Subspecies: (1) Hoffmann's Checkerspot (*C. h. hoffmanni*) (Pl. 6k), described above; Sierra Nevada of Cali-fornia and western Nevada. (2) Segregated Checkerspot (*C. h. segregata*), darker and more heavily marked; Siskiyou County north into Oregon. Fairly common in the Mt. Shasta area. TL Crater Lake, Oregon. *Early stages*: *eggs* light green, laid in masses; *larva* black, spiny, with scalloped cream-colored lat-eral line and brown sides; *pupa* white to brown with dark mot-tling. *Larval food plant*: aster (*Aster conspicuus*) (Asteraceae) in Washington; unknown for California.

Crocale Patch (*Chlosyne lacinia crocale*) (Pl. 6l). The most highly polymorphic of our butterflies, the Crocale Patch occurs in three previously named forms: nominate *crocale*, black with a white band; form "*rufescens*," the band yellow to rusty red; form "*nigrescens*," the light band greatly reduced or lacking. Seldom do two specimens look exactly alike. The Crocale Patch is found in cultivated areas of the Imperial and Coachella valleys, with strays in mountains of Riverside and San Diego counties. Flight period March–October; several broods. Common at Blythe in September. *Early stages*: *egg* greenish yellow; *larva* (the overwintering stage) black, orange, or black-and-orange striped, with black spines; *pupa* black or white. *Larval food plant*: Common Sunflower (*Helianthus annuus* var. *lenticularis*) (Asteraceae).

California Patch   (*Chlosyne californica*) (Pl. 6e). A tawny patch, as befits the desert areas in which it flies. The basal area and submarginal bands are dark brown, separated by a wide tawny band. Found in canyons and washes bordering the Coachella and Imperial valleys, as at Whitewater and Chino canyons, Riverside County, and Sentenac Canyon, San Diego County, east through Mojave and Colorado desert ranges. Flight period March–April, and again in June and September, depending on rainfall; several broods. *Early stages*: *egg* yellow green; *larva* black with branching spines and rows of white dots; *pupa* white with black splotches on wing cases. *Larval food plant*: Desert Sunflower (*Viguiera deltoides* var. *parishii*) (Asteraceae).

Leanira Checkerspot   (*Thessalia leanira*). Usually uncommon; occurs in local colonies. A small but striking species. UP black with narrow white bands. UNHW cream colored; a black band, enclosing small white spots, crosses both wings. Usually a foothill species, flying around rocky hills and outcrops. Flight period April–May (south), May–June (north); one brood. Five subspecies are found in California: (1) Leanira Checkerspot (*T. l. leanira*) (Pl. 7g), Coast Ranges and northern central California. (2) Davies's Checkerspot (*T. l. daviesi*), UP white bands wider; foothills of Sierra Nevada. (3) Wright's Checkerspot (*T. l. wrighti*), dark, with red-and-white markings; canyons of Los Angeles and Orange counties, also western San Bernardino County, where it intergrades with desert *cerrita*. (4) Cerrita Checkerspot (*T. l. cerrita*) (Pl. 7h), UP broadly reddish tawny, dark only near the wing bases; deserts of southern California, north at least to eastern Mono County. (5) Alma Checkerspot (*T. l. alma*), pale tawny to yellowish, with most dark markings suppressed; Panamint Mountains, Death Valley National Monument, and an indefinite distance eastward. *Early stages*: *eggs* laid in masses on food plant; *larva* black, spiny, with two dorsal and two lateral lines of small orange spots. *Larval food plants*: paintbrush (*Castilleja* spp.), bird beak (*Cordylanthus* spp.) also recorded (Scrophulariaceae).

Imperial Checkerspot   (*Dymasia chara imperialis*) (Pl. 7f). Named for the valley in which it flies, not for its regal status, the Imperial Checkerspot is the most diminutive of our checkerspots, and might go unnoticed by any but a skilled observer. Attracted, as are hummingbirds, to the Chuparosa, on the nectar of which both feed, it occurs on the margins of the Coachella and Imperial valleys of Riverside and Imperial counties, and in one coastal area of San Diego County. Flight period March–April and also September–October after rains; two broods. *Early stages*: *egg* light yellow; *larva* gray with black-and-white mottling and orange spots at bases of spines; *pupa* gray with black or brown speckles. *Larval food plant*: Chuparosa (*Beloperone californica*) (Acanthaceae).

Monache Checkerspot   (*Poladryas arachne monache*) (Pl. 5l). UP uniform tawny, irregularly striped and spotted with blackish. UNFW ochreous with yellowish white near tip; UNHW with four bands of creamy white, the intervening dark bands continuous or nearly so. High valleys of Tulare County, south of Mt. Whitney, such as the Monache Meadows, for which it was named. Flight period late June and early August; possibly two broods. Nearest relatives occur in Arizona, Colorado, and New Mexico. *Early stages*: little known. *Larval food plant*: beard tongue (*Penstemon* spp.) (Scrophulariaceae), abundant on the dry slopes on which *monache* flies.

Distinct Crescent   (*Phyciodes pascoensis distinctus*) (Pl. 7l). Crescentspots ("Crescents") resemble small checkerspots, with UN more uniformly colored and with a pale crescentic marking on outer edge UNHW. The Northern Pearl Crescent ranges across the northern tier of states, invading the southwest via Texas, New Mexico, and Arizona, where subspecies *distinctus* occurs. In California it is found only in the Imperial Valley near Brawley, where it flies along irrigation ditches and in well-watered places in late summer and fall.

Several broods in East; probably two or more in California. *Early stages*: *eggs* laid in clusters; *larva* black with brownish spines and yellow dots; *pupa* gray or brown. *Larval food plants*: unknown for California; elsewhere, asters (*Aster* spp.) (Asteraceae).

Phaon Crescent (*Phyciodes phaon*) (Pl. 7m). A small, dark crescent with short, rounded FW, tawny color restricted to central area of wings, small creamy white patches on UPFW only. In California limited to coastal San Diego and inland Riverside and Imperial counties, as at Santa Fe, Blythe, and Lakeside. Early reports from San Bernardino lack confirmation. Flight period March–October; several broods. *Early stages*: *eggs* laid in small groups; *larva* olive, with lengthwise lateral brown band and scalloped white band below spiracles; *pupa* brown, speckled with black and white. *Larval food plants*: Fog Fruit (*Lippia lanceolata*) and Mat Grass (*L. nodiflora* var. *rosea*) (Verbenaceae).

Field Crescent (*Phyciodes campestris*). A dark species, with heavy black markings. It frequents fields, meadows, and fence rows, flying low over the ground. Flight season spring to fall at low elevations; one generation in midsummer at high elevations. Subspecies: (1) Field Crescent (*P. c. campestris*) (Pl. 7j), as noted above; cismontane California and north to Alaska. (2) Mountain Crescent (*P. c. montanus*) (Pl. 7i), much brighter, with fewer dark markings, especially UPHW; moderate to high elevations in the Sierra Nevada. *Early stages*: *eggs* laid in a cluster; *larva* spiny, black, a light dorsolateral line, underside brown; *pupa* wood brown with a network of dark brown lines. *Larval food plants*: in southern California, Marsh Aster (*Aster hesperius*); in central and northern California, California Aster (*Aster californicus*), and others (all Asteraceae).

Mylitta Crescent (*Phyciodes mylitta*) (Pl. 7k). Bright tawny with narrow dark markings, in contrast to the dark Field Crescent, with which it often flies. Flight season early spring until late fall at lower elevations, where it is several brooded;

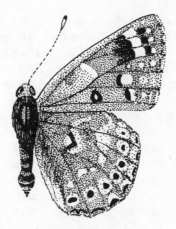

FIG. 26 Orseis Crescent

midsummer at higher elevations, where there is probably only one brood. Found throughout California and the West, north into Canada. *Early stages*: *egg* white; *larva* spiny, black, with yellow hair tufts on some segments; *pupa* gray with a slight sheen. *Larval food plants*: various thistles (*Cirsium* spp.), Milk Thistle (*Silybum marianum*) (Asteraceae).

Orseis Crescent   (*Phyciodes orseis*) (Figure 26). As large as, or larger than, the Field Crescent. Somewhat resembles the Mylitta Crescent; color deeper, reddish rather than tawny. Inner Coast Ranges in Napa, Sonoma, Mendocino counties, northward as far as Diamond Lake, Oregon. The coastal populations are scarce and local; not taken in some places for many years. A montane population, Herlan's Crescent (*P. o. herlani*), more tawny, less reddish, more lightly marked, has recently (1975) been discovered near Lake Tahoe, flying in the central Sierra Nevada in midsummer. Flight period April–June; one brood. *Early stages*: *egg* ribbed, greenish; *larva* dark and spiny; *pupa* mottled brown with tubercles. *Larval food plants*: thistles (*Cirsium* spp.) (Asteraceae).

Satyr Anglewing   (*Polygonia satyrus neomarsyas*) (Pl. 8d). A glance at these insects shows how well they deserve the

popular name *anglewings*. The outer edges of all four wings
are incised with irregular indentation. All anglewings are
rather similar: UP rusty red, UN colored like dried leaves or
dead bark, and silver comma or *V* in center of UNHW. The
Satyr is recognized by the lighter colored UP and the browner
UN. All anglewings live a long time as adults, and pass the
winter in this stage. They often fly on warm winter days. Flight
period April–November; two to three broods. *Early stages*:
egg rounded, taller than wide, with about a dozen flangelike
ribs, pale green; *larva* covered with scattered spines, dark with
a greenish white stripe down the back; *pupa* wood brown,
rough, covered with tubercles (small bumps). *Larval food
plant*: Hoary Nettle, "Stinging Nettle," (*Urtica holosericea*)
(Urticaceae).

Rustic Anglewing   (*Polygonia faunus rusticus*) (Pl. 8c).
A western subspecies of a species found in many parts of the
U.S. and Canada. UP dark reddish brown; UN with a greenish
gray tint and small greenish spots just inside the outer edge.
Flight period late May–July; single brood. *Early stages*: *egg*
undescribed; *larva* reddish tawny, with a dorsal light area;
spines white; *pupa* light brown with metallic spots. *Larval
food plants*: of the species: azalea (Ericaceae), gooseberry
(Saxifragaceae), birch, alder (Betulaceae), and willow (Salicaceae). Our western subspecies, found from Santa Cruz
County and the central Sierra Nevada north to Canada, favors
streamsides, where it is most often associated with Western
Azalea (*Rhododendron occidentale*) (Ericaceae).

Oreas Anglewing   (*Polygonia oreas*) (Pl. 8b). Smaller
than our other anglewings; UP a rich reddish tint; dark markings small and widely spaced, UN deep lavender gray, the basal
half darker; UN etched with fine light cross-hatchings; silver V
thin and bright. Local and uncommon, usually found in the vicinity of streams, from Monterey County north. This is the *P.
o. silenus* of Comstock. The name *silenus* applies to the northern subspecies, not found in California. Flight period early
spring to late fall; two broods: midsummer and fall. *Early
stages*: unknown. *Larval food plant*: Straggly Gooseberry
(*Ribes divaricatum*) (Saxifragaceae).

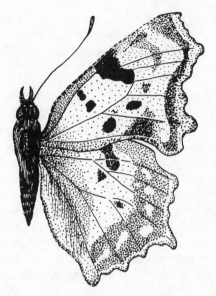

FIG. 27 Sylvan Anglewing

Sylvan Anglewing (*Polygonia silvius*) (Figure 27). UN very dark. Rare and little known. Records believed to apply to this species are from the Coast Ranges and the Sierra Nevada at moderately high elevations in midsummer. A population has been found recently in Sonoma County. Flight period May–August; one brood. *Early stages*: *egg* pale green; *larva* black, fulvous, and white, spined; *pupa* tan, marbled, olive, and brown. *Larval food plant*: Western Azalea (*Rhododendron occidentale*) (Ericaceae).

Zephyr Anglewing (*Polygonia zephyrus*) (Pl. 8a). Somewhat resembles the Oreas Anglewing; larger, not so dark. UPFW submarginal pale spots with small dark centers, usually a good diagnostic feature. UN basally dark gray and brown, outwardly lighter, with many tiny pale flecks. Most common from the Sierra-Cascade crest eastward; less common in high mountain ranges of southern California. Found in open forest and along streams in the sagebrush (Artemisian) habitat. Flight period late June–September; one brood. *Early stages*: *egg* unrecorded; *larva* black with yellow spots above, and branching spines; *pupa* olive and salmon pink with silver spots. *Larval food plants*: currant and gooseberry (*Ribes* spp.) (Saxifragaceae).

California Tortoise Shell   (*Nymphalis californica*) (Pl. 8e). This species is subject to wide fluctuations in numbers. In some years great outbreaks occur. In other years it is a scarce butterfly. In years of abundance these butterflies swarm across valleys and through the foothills, into the higher mountains, by millions, covering roads and plastering windshields and radiators. Although this species is more common in central and northern California, migrations do occur in the higher ranges of southern California. Overcrowding has been a usual explanation for these altitudinal migrations. But Shapiro has suggested that the food plants become high in tannin as the season advances, and that the adults go higher into the mountains to find new growth that is low in tannin. Flight period April–June for overwintering adults, June–August for later broods; one to three broods. *Early stages*: *egg* unrecorded; *larva* black, with a blue tubercle at the base of each spine and an indistinct pale line or shade down the back; some individuals are lighter; *pupa* gray to pale brown with darker markings, the surface often rough or slightly spiny. *Larval food plants*: *Ceanothus* spp., including Blue Blossom (*C. thyrsiflorus*), Buck Brush (*C. cuneatus*), Snow Brush (*C. cordulatus*), White Thorn (*C. incanus*), and Deer Brush (*C. integerrimus*) (Rhamnaceae).

Mourning Cloak   (*Nymphalis antiopa*) (Pl. 23c). Few butterflies are better known or have a wider range than this distinctive species. It is found in Europe and Asia as well as in North America. It is rare in England, where it is called the Camberwell Beauty. The dark brown color and yellow edge, lined with blue spots, are unique. When its wings are closed, it resembles a dead leaf, as do anglewings and tortoise shells. It is longlived, as are many nymphalids. The adults seek shelter and overwinter, appearing on the first warm days of late winter and early spring. Flight season early spring to late fall; one brood in early summer; a second brood in some areas. *Early stages*: *egg* pale green fading to whitish, turning dark before hatching; *larva* spiny, purplish black, a double row of red spots at spine bases along the back; *pupa* purplish gray, wing pads brownish, abdomen spiny. *Larval food plants*: willow (*Salix* spp.), cottonwood (*Populus* spp.) (Salicaceae); elm (*Ulmus* spp.) (Ulmaceae); possibly others. The Mourning Cloak is an easy species to rear, and makes a fine insect for classroom study.

Milbert's Tortoise Shell   (*Aglais milberti furcillata*) (formerly *Nymphalis*) (Pl. 8f). Smaller than the California Tortoise Shell; UP dark brown, a wide orange brown band near outer edge of wings; inner edge of this band shaded with yellow in the western subspecies. A striking butterfly, uncommon in the coastal regions, but more common at high elevations in the Sierra Nevada and Cascade Range, and in the higher mountains of southern California. Overwintering adults may be found in early spring. Flight period for fresh adults May–September; probably two broods in California. *Early stages*: *egg* conical, green, with vertical ribs; *larva* black with branching spines and white dots, a yellow line lengthwise and a bright orange stripe; *pupa* light gray to black, spiny. *Larval food plant*: Hoary Nettle (*Urtica holosericea*) (Urticaceae).

Red Admiral, Alderman Butterfly   (*Vanessa atalanta*) (Pl. 8k). Not to be confused with the admirals of the genus *Basilarchia*, this butterfly, its black ground color crossed by a bright red band, is like none other we have. The red flashes brightly in flight, but the dark underside, with its neat small eyespots, tends to conceal the butterfly at rest. The Red Admiral is found in Europe and Asia as well as in North America, where subspecies *rubria* occurs. It is not very common in much of California, and is quite a prize for junior collectors. Adults overwinter, may be active on warm winter days, returning often to the same perch, or series of backyard perches. Flight period much of the year; at least three broods. *Early stages*: *egg* green, barrel shaped, with about ten vertical flanges; *larva* black, spiny; lives in a shelter made by drawing the edges of a leaf together with silken threads; *pupa* gray with gold tubercles (lumps). *Larval food plants*: Hoary Nettle (*Urtica holosericea*) and other nettles (Urticaceae); also hops (*Humulus*) (Moraceae).

Painted Lady, Thistle Butterfly   (*Vanessa cardui*) (Pl. 8i). Said to be the most widely distributed butterfly in the world, the Painted Lady has been taken by man's activities to many places far removed from the temperate parts of the Northern Hemisphere to which it is native. Like the Monarch, it is a migrant, but unlike the Monarch, it does not have return flights. In the spring of 1952 and 1958, for example, millions of individuals swept northwesterly across the deserts of southern California into central California and north to Oregon. They fly a few feet above the ground, rising vertically over obstacles, rather than going around them. During the rest of the year, they may be found in almost any vacant lot or backyard. The Painted Lady is distinguished from the American Painted Lady by the row of small eyespots UNHW, and from the West Coast Lady by the white bar UPFW about two-thirds the distance from wing base to wing tip. Adults overwinter. Flight season most of year; several broods. *Early stages*: *egg* green, barrel shaped, with fourteen or more vertical ribs; *larva* lavender to pale brown, with two yellowish lateral lines and dark lines below these; spiny; lives in a shelter made by drawing leaves together with silk threads; *pupa* brown with black dots and golden spots. *Larval food plants*: various thistles (*Cirsium* spp.); Italian Thistle (*Carduus pycnocephalus*) (Asteraceae); nievitas (*Cryptantha* spp.) and fiddleneck (*Amsinckia* spp.) (Boraginaceae); sometimes nettle (*Urtica* spp.) (Urticaceae); mallow (*Malva* spp.) (Malvaceae); and others.

American Painted Lady, Virginia Lady   (*Vanessa virginiensis*) (Pl. 8g). This pretty species is native to the New World, where it is generally distributed. It may be distinguished from either the Painted Lady or the West Coast Lady by the two large eyespots UNHW. Less common than these, it is found throughout coastal and montane California, rarely in desert areas. Several broods a year; adults every month in southern counties. *Early stages*: *egg* greenish, barrel shaped, ribbed; *larva* lavender, with black patches surrounding black spines and enclosing paired white spots; *pupa* light gray to brown with darker lines on abdomen. *Larval food plants*: cudweed (*Gnaphalium* spp.), Pearly Everlasting (*Anaphalis margaritacea*), and at times other composites (Asteraceae).

West Coast Lady   (*Vanessa annabella*) (Pl. 8h). Previously mistakenly called *carye*, a name that belongs to a species native to Chile. The West Coast Lady is common at low elevations in California, where adults may be found nearly all year. Distinguished from other ladies by the orange bar UPFW about two-thirds the distance from wing base to tip, and by the square-cut wing tips. Several broods a year. An easy species to rear in home or classroom. *Early stages*: *egg* similar to that of American Painted Lady; *larva* variable in color, from tan through brown to black with yellow lines, spiny, lives in a leaf shelter; *pupa* brown with golden flecks. *Larval food plants*: members of the Mallow Family (Malvaceae), including mallow (*Malva* spp.), Hollyhock (*Althea rosea*), Checkerbloom (*Sidalcea malvaeflora*), globe mallow (*Sphaeralcea* spp.), Tree Mallow (*Lavatera assurgentiflora*), and others; nettle (*Urtica* spp.) (Urticaceae).

Buckeye, Peacock Butterfly   (*Junonia coenia*) (Pl. 8l). The light brown ground color and the large multicolored (peacock) eyespots make a pattern that once seen will be remembered. The Buckeye is common in California, flying from spring to fall; two or more broods. Late fall specimens may be reddish or purplish UNHW. Old neglected fields and weedy pastures are good places to find Buckeyes; so are roadsides and city yards, at times. Males may patrol a "territory," flying out to investigate every large insect that passes. *Early stages*: *egg* roundish, wider than high, dark green, ribbed; *larva* black, often with two yellow or tawny stripes, very spiny; *pupa* brown, more curved than that of *Vanessa*. *Larval food plants*: Common Plantain (*Plantago major*), Ribwort (*P. lanceolata*), and others (Plantaginaceae); Common Snapdragon (*Antirrhinum majus*), of which it may become a garden pest, and monkey flower (*Mimulus* spp.), including Musk Flower (*M. moschatus*) (Scrophulariaceae).

Dark Buckeye, Dark Peacock   (*Junonia nigrosuffusa*) (Pl. 8j). An invader from the American tropics via Texas, New Mexico, and Arizona, the Dark Buckeye crosses the Colorado River at Blythe in eastern Riverside County. Two- or three-brooded elsewhere, it is found in California only in the fall. As the name implies, it is darker than the Buckeye, the ground color being brownish black; the eyespots UPHW are relatively small; the UNHW has a dark transverse band. *Early stages*: unrecorded. *Larval food plants*: in Texas, Wooly Stemodia (*Stemodia tomentosa*); in Arizona and California, possibly *S. durantifolia* (both Scrophulariaceae).

Arizona Viceroy   (*Basilarchia archippus obsoleta*) (formerly *Limenitis*) (Pl. 3c). A rarity in California, this western member of the regal trio, Monarch, Queen, and Viceroy, flies only in the hot and humid Colorado and Imperial valleys. Unlike the other two, however, it is not a danaid, but a nymphalid, and an admiral at that. This is proven, not by the adult, which mimics the distasteful Queen, but by the caterpillar and chrysalis, which are of true *Basilarchia* form. Flight period April–October; several broods. *Early stages*: *egg* ovoid, finely hairy, pitted; *larva* brownish or grayish, a white dorsal patch, and two plumelike horns arising from the thorax; *pupa* with brownish wing covers, abdomen whitish, a conspicuous brown outgrowth or bulge. *Larval food plants*: cottonwood (*Populus* spp.) and willow (*Salix* spp.) (Salicaceae).

Wide-banded Admiral   (*Basilarchia weidemeyerii latifascia*) (formerly *Limenitis*) (Pl. 23d). A large, showy species, black, the wings crossed by a pure white band. *Latifascia* is found east of the Sierra Nevada. Other subspecies of Weidemeyer's Admiral are found eastward through the mountains of Nevada, Utah, and Colorado, to the Black Hills of South Dakota, usually around aspen groves. Flight period late June–early August; one brood. *Early stages*: *egg* nearly round, surface pitted with hexagonal cells; *larva* of *Basilarchia* form, grayish, with dark gray and whitish mottling; *pupa* with prominent rounded protuberance on mid-dorsum. *Larval food plants*: most often Quaking Aspen (*Populus tremuloides*); poplar and cottonwood (*Populus* spp.), also willow (*Salix* spp.) (Salicaceae). A hybrid between this species and Lorquin's Admiral, which has been called Friday's Admiral (*B. weidemeyerii latifascia* × *B. lorquini*), is found in the vicinity of Mono Lake, where these two species meet. This hybrid looks most like the Wide-banded Admiral, but has a suggestion of the red tips and the brown UN markings of Lorquin's Admiral. It is a rarity, and quite local.

Lorquin's Admiral (*Basilarchia lorquini*) (formerly *Limenitis*) (Pl. 23f). The name of this butterfly honors Pierre Lorquin, the French collector of the Gold Rush days of California who discovered this and many other species of California butterflies. Lorquin's Admiral is a showy insect, with velvety black ground color, pure white band, and brick red wing tips. It flies with a few quick wing beats alternated with gliding. It frequents stream courses and moist meadows in the foothills and mountains, and flies along valley streams, from late spring to fall. It overwinters as a larva in a case made from a leaf. Two or three broods at low elevations, one at high elevations. *Early stages*: *egg* silvery green, nearly spherical, with raised network and short spines; *larva* brown, the head bilobed, two prominent horns behind head, and various bumps along the back; *pupa* purplish brown with light markings, the wing covers olive, a big hump in the middle of its back. *Larval food plants*: willow (*Salix* spp.), cottonwood and poplar (*Populus* spp.) (Salicaceae); Western Choke Cherry (*Prunus virginiana* var. *demissa*) (Rosaceae); sometimes orchard trees such as prune, cherry, plum, and apple (Rosaceae).

California Sister (*Adelpha bredowii californica*) (formerly *Limenitis*) (Pls. 1i, 1j, 3g, 23e). A regal species, stately in flight. Often seen around forest trees, especially oaks. The color pattern is similar to that of Lorquin's Admiral, but the Sister is larger, with orange (not brick red) wing tips bordered by the ground color (not bordered in Lorquin's Admiral), and with broad bluish shadings UN. A "flyway" butterfly that patrols woodland paths and often returns to a favorite perch. It seldom visits flowers, but is attracted to water. Found throughout the Coast Ranges, and in the oak belt of the Transverse, Sierra Nevada,, and Cascade ranges. Flight period March–September, shorter at high elevations; two or three broods. The caterpillar overwinters. In the eastern Mojave Desert ranges of San Bernardino County, the Arizona Sister (*A. b. eulalia*) flies from May to July, there being but a single brood. *Early stages*: egg nearly hemispherical, green; *larva* similar in shape to that of Lorquin's Admiral, dark green; *pupa* brown with a smaller hump than that of Lorquin's Admiral, and with two sharp lateral projections from the head. *Larval food plants*: various oaks, especially Canyon Oak (*Quercus chrysolepis*) and Coast Live Oak (*Q. agrifolia*); in central Sierra Nevada often associated with Huckleberry Oak (*Q. vaccinifolia*) (all Fagaceae).

Swallowtails and Parnassians (Family Papilionidae)

A family comprising two groups of medium-sized to large butterflies that, although closely related, do not look very much alike.

Swallowtails: large butterflies, 2¼–5 in. (50–125 mm), of powerful flight, and (in California) with tails on the hind wings. Colors predominantly black and yellow, with accents of blue or red; *adult* with anal margin of HW folded under abdomen; antennae long and curved. *Egg* nearly spherical. *Larva* with osmateria (scent horns). *Pupa* suspended by both cremaster and silken girdle (as in Pieridae), but wing cases not enlarged, and head usually notched, not long and pointed. *Larval food plants*: members of the carrot family (Apiaceae), broad-leaved trees and shrubs, or Pipe-vine (*Aristolochia californica*) (Aristolochiaceae). Found from desert regions to mountain tops.

Parnassians: medium-sized butterflies, 1¾–2½ in. (44–62 mm); HW evenly rounded, without tails; color chalky white or semitransparent, usually with a few red or yellow spots. Females acquire an abdominal sac (sphragis) during copulation that hardens to prevent other matings. *Egg* turban-shaped, slightly roughened. *Larvae* scent horns poorly developed. *Pupa* in cocoon on ground. *Larval food plants*: stonecrop (*Sedum* spp.) (Crassulaceae); bleeding heart (*Dicentra* spp.) (Fumariaceae). Found in upper middle elevations and at high altitudes; also in far north. Two species in California.

Clodius Parnassian   (*Parnassius clodius*). The Clodius Parnassian has black antennae, thin and somewhat translucent wings, and no red (or yellow) spots in FW. Four rather similar subspecies are found in California, all north of Kern County (there are no parnassians in southern California, although montane habitats appear suitable): (1) Clodius Parnassian (*P. c. clodius*) (Pls. 10e, 23b), Coast Ranges from Marin County north. (2) Sol Parnassian (*P. c. sol*), west slopes of Sierra Nevada–Cascade Range in northern California. (3) Strohbeen's Parnassian (*P. c. strohbeeni*), Santa Cruz Mountains; this population now seems to be extinct; it was last seen in 1958. (4) Baldur Parnassian (*P. c. baldur*), moderate to high elevations in the Sierra Nevada. Flight period (for the species) June–August, depending on elevation; one brood. *Early stages*: much like those of the Phoebus Parnassian (see below). *Larval food plants*: Bleeding Heart (*Dicentra formosa*); in the Trinity Alps, Shorthorn Steershead (*Dicentra pausiflora*) (Fumariaceae).

Small Apollo, Phoebus Parnassian   (*Parnassius phoebus*), formerly known as the Smintheus Parnassian (*Parnassius smintheus*). Usually smaller than the Clodius Parnassian. Antennae ringed black and white; wings more opaque; FW has two red (or yellow) spots, these more evident in females. A truly alpine species, favoring rocky summits. One brood, midsummer. Two California subspecies: (1) Sternitzky's Parnassian (*P. p. sternitzkyi*), high mountains of Siskiyou County. (2) Behr's Parnassian (*P. p. behrii*) (Pls. 9e, 23a), very high elevations in the central Sierra Nevada, from Tulare/Inyo counties (Mt. Whitney) to at least Plumas County. *Early stages: larva* has small patches of hairs, black with yellow spots (resembling certain millipedes); *pupa* plump, brown, in cocoon on ground. *Larval food plants*: various species of stonecrop (*Sedum* spp.) (Crassulaceae).

Pipe-vine Swallowtail   (*Battus philenor*). No other large butterfly found in California is iridescent greenish black UP, with red spots UNHW. A conspicuous species, not easily over-looked. Flight period spring and summer; two broods a year. Two California subspecies: (1) Pipe-vine Swallowtail (*B. p. philenor*) (Pl. 9d), found over much of the United States, oc-curs as a stray in southern California, often near the coast. Garth has taken it in Long Beach. (2) Hairy Pipe-vine Swal-lowtail (*B. p. hirsutus*), smaller, shorter tails, body hairy; from San Mateo and Alameda counties north to Shasta County, in Coast Ranges and Sierra Nevada foothills as well as northern Sacramento Valley. *Early stages*: *egg* spherical, rather rough, reddish brown; *larva* black with bright red spots and long, fleshy filaments; *pupa* either dark brown or green, somewhat S-shaped in side view. *Larval food plant*: Dutchman's Pipe (*Aristolochia californica*) (Aristolochiaceae), a vine with roundish dark green leaves and brown pipe-shaped flowers.

Baird's Swallowtail   (*Papilio bairdii*) (Pl. 9a). A large, predominantly black swallowtail with a narrow spot band across the wings; female almost entirely black with blue spots UPHW. Found in California only in the San Bernardino Moun-tains, where it flies from the bottom of the Santa Ana River Canyon to the top of Mt. San Gorgonio, el. 11,500 ft. Flight period April–September; several brooded, depending on rain-fall. Polymorphic (many forms). Forms that have received names are Bruce's Swallowtail (form "*brucei*"), with more yellow; and Holland's Swallowtail (form "*hollandi*"), yellow on UPFW, UPHW with more black. *Early stages*: *egg* spheri-cal, light green; *larva* green with black bands, these bands with orange spots; *pupa* brown or green. *Larval food plants*: Dragon Sagewort, Tarragon (*Artemisia dracunculus*) (As-teraceae); may be reared on Sweet Fennel, Anise (*Foeniculum vulgare*) (Apiaceae).

Anise Swallowtail, Western Parsley Swallowtail   (*Papilio zelicaon*) (Pl. 10b). A deep yellow swallowtail, with rather wide black bands. It flies in vacant lots and along roadsides, as well as in hills and fields, throughout the state except in desert regions, where it is replaced by Wright's Swallowtail, which resembles it but is smaller and a bit duller. Flight period February–October; several broods. *Early stages*: *egg* smooth, spherical, pale green; *larva* black with orange spots when young, greenish with dark bands when mature; protrudes osmateria (scent horns) from just back of head when disturbed; *pupa* either green or brown. *Larval food plants*: originally native members of the Parsley family (Apiaceae), such as tauschia (*Tauschia* spp.), lomatium (*Lomatium* spp.), pteryxia (*Pteryxia* spp.); now usually associated with Anise (Sweet Fennel) (*Foeniculum vulgare*), also Carrot (*Daucus carota*) (Apiaceae). It is often easier to rear the larvae than to catch the adults!

Wright's Swallowtail (*Papilio polyxenes coloro*) (Pl. 10f). Until October 1982, known under its synonymic name, *P. rudkini*. The name change from *rudkini* to *coloro* was required when it was discovered that a specimen of W. G. Wright's collecting in the California Academy of Sciences, long thought to be an aberrant *zelicaon*, had the wider outer black wing bands associated with the species that had meanwhile come to be called *rudkini*, first described by J. A. Comstock as an aberration, but raised to species level by F. and R. Chermock. As a name for the subspecies, Wright's *coloro*, TL "Colorado Desert of southern California," antedates Comstock's *rudkini*, and so correctly applies. And although common names are not subject to the rules of zoological nomenclature, it seems to your authors that what was once called Rudkin's Swallowtail should now be called Wright's. A medium-sized yellow-and-black swallowtail of the deserts of southern California, from Victorville to Death Valley, east into Nevada, and from Whitewater, Riverside County, to Coachella Valley, east into Arizona. Polymorphic. Two forms have received names; Clark's Swallowtail (form "*clarki*"), mostly black; and Comstock's Swallowtail (form "*comstocki*"), UPHW bluespotted. Flight period February–October; several broods. Abundance depends on rainfall; summer rains trigger emergences. *Early stages*: *egg* spherical, green; *larva* variable, pale green or white with black bands, to black with light bands, each form with yellow dots; *pupa* brown or green. *Larval food plants*: Turpentine Broom (*Thamnosma montana*) (Rutaceae); may be reared on Sweet Fennel or Anise (*Foeniculum vulgare*) (Apiaceae). A rapid flier, Wright's Swallowtail may be obtained in good numbers by gathering larvae from the food plant a month in advance of the flight period.

Indra Swallowtail   (*Papilio indra*). An elegant and usually wary species. The males are attracted to moist places, where they are more easily taken. California has six recognized subspecies: (1) Short-tailed Swallowtail (*P. i. indra*) (Pl. 9b), small, black with narrow yellow band across the wings, and very short tails; northern California in Coast Ranges and Sierra Nevada. (2) Edwards's Swallowtail (*P. i. pergamus*) (Pl. 9f), larger, long tailed, yellow band narrow; San Gabriel, San Bernardino, Riverside, and San Diego counties. (3) Ford's Swallowtail (*P. i. fordi*), smaller, yellow band wide; western Mojave Desert, Granite Mountains and associated ranges, San Bernardino County. (4) Martin's Swallowtail (*P. i. martini*), FW yellow band narrow, HW band tending to become narrow or disappear; Providence Mountains, eastern San Bernardino County. (5) Phyllis's Swallowtail (*P. i. phyllisae*), median band wide; TL Butterbread Peak, Kern County; Piute Mountains north on east side of Sierra Nevada to Whitney Portal. (6) Panamint Swallowtail (*P. i. panamintensis*), median band UPFW moderately wide, HW band narrow, only on upper edge of HW; TL Thorndike Campground, Wildrose Canyon, Panamint Range; Cottonwood Mountains, and Last Chance Range, Inyo County. Flight period differs with different subspecies; March–June (July). Basically one brooded; some populations have a smaller second brood, at least in some years. *Early stages*: egg cream colored; *larva* banded black with yellow or orange; *pupa* light tan to brown, may be greenish or grayish. *Larval food plants*: many plants of the Parsley Family (Apiaceae). In central and northern Sierra Nevada, Terebrinth Pteryxia (*Pteryxia terebrinthina*); in southern Sierra Nevada, Parish's Tauschia (*Tauschia parishii*); in coastal mountains of southern California, Southern Tauschia (*T. ar-*

guta) and Shiny Lomatium (*Lomatium lucidum*); in the Panamint and Providence mountains, Parry's Lomatium (*L. parryi*); in the Granite Mountains, Panamint Cymopterus (*Cymopterus panamintensis* var. *acutifolius*).

Giant Swallowtail   (*Papilio cresphontes*) (Pl. 9c). A large species with a bold pattern of yellow and black, including a yellow spot on each wide tail. Ranging through the central, eastern, and southern U.S. into the tropics, the Giant Swallowtail is a recent invader of the citrus-growing areas of our state. First found in California in the Imperial Valley, it has spread since 1963 to the vicinity of Blythe, Riverside County, and to the San Joaquin Valley near Porterville, Tulare County. Flight period March–October. *Early stages*: *egg* spherical, light green; *larva* dark brown, light marks on thorax, light "saddle" and markings near anal end (called the "Orange Dog" in eastern and southern U.S.); *pupa* brown. *Larval food plants*: citrus trees, such as orange, lemon, and grapefruit (Rutaceae) in California. In other parts of its range many other food plants are used.

Western Tiger Swallowtail   (*Papilio rutulus*) (Pls. 1o, 1p, 10c). The bright yellow and black of the Western Tiger is a familiar sight throughout California, as it flies high over streets or city parks or visits backyard flowers. In uncultivated areas it frequents watercourses where grow the trees that are its larval food plants. It belongs to the riparian (streamside) community rather than to the drier slopes that the Pale Swallowtail usually inhabits. Flight period early spring to midsummer, in some places to late fall; one to three broods, depending on elevation. *Early stages*: *egg* spherical, green; *larva* bright green, with big "false eyes" on the third thoracic segment; these may function to scare predators; there is a black-and-yellow bar just behind the "eyes"; the larva lives in a shelter made by drawing leaves together with silken threads; *pupa* dark brown. *Larval food plants*: many deciduous broad-leaved trees; cottonwoods (*Populus* spp.), Quaking Aspen (*Populus tremuloides*), various willows (*Salix* spp.) (Salicaceae); White Alder (*Alnus rhom-*

bifolia), Red Alder (*A. rubra*), Water Birch (*Betula fontinalis*) (Betulaceae), Western Sycamore (*Platanus racemosa*) (Platanaceae); elms (*Ulmus* spp.) (Ulmaceae); also orchard trees, such as Apple (*Malus malus*) (Rosaceae).

Pale Swallowtail   (*Papilio eurymedon*) (Pl. 10a). This elegant species suggests a paler Western Tiger, but close inspection shows the more pointed forewings, the wider black borders, and the long tail with a half-twist. A butterfly of the hills, mountains, and canyons, it is on the wing from March to September in some localities; two broods. As with our other swallowtails, the spring adults emerge from overwintering chrysalids. Males, in common with other male swallowtails, are "hilltoppers," often flying around the summits of exposed hills. The Pale Swallowtail, like other swallowtails, is attracted to water as well as to flowers, and numbers may sometimes be found on moist sandbars, their heads all facing the same way. *Early stages*: *egg* spherical, yellow green or pinkish; *larva* soft green, the "false eyes" reduced to a few dark markings; *pupa* dark brown. *Larval food plants*: buckthorn (*Rhamnus* spp.), including California Coffee Berry (*R. californica*) and Redberry (*R. crocea*); *Ceanothus* spp., including Buck Brush (*C. cuneatus*) and Deer Brush (*C. integerrimus*) (all Rhamnaceae); Holly-leaved Cherry (*Prunus ilicifolia*) (Rosaceae).

Two-tailed Swallowtail   (*Papilio multicaudatus*) (Pl. 10d). Largest of our swallowtails, superficially similar to the Western Tiger Swallowtail, this handsome species has two tails on each hind wing. The black bands are narrow, so that there is much more yellow than black. The Two-tailed Swallowtail is found, often sparingly, throughout northern California. In southern California it flies only in the drier mountain ranges, such as the Tehachapi and San Gabriel mountains on the desert-facing side, and in desert ranges, such as the Panamint and Providence mountains. Flight period June–August; seldom common; two or more broods. *Early stages*: much like those of Western Tiger Swallowtail, but the "eyes" of the larva are much smaller. *Larval food plants*: Western Choke Cherry

(*Prunus virginiana* var. *demissa*) (Rosaceae); Hop Tree (*Ptelea crenulata*) (Rutaceae); ash (*Fraxinus* spp.) (Oleaceae).

Whites, Sulfurs, Marbles, and Orange-tips (Family Pieridae)

Moderate-sized [¾–2¾ in. (19–69 mm)] yellow or white butterflies with dark markings. Legs fully developed; radial and medial vein branches of FW closely associated. Some wing colors formed from uric acid deposits. Some of the sulfurs are deep orange; some white butterflies have orange wing tips and/or delicate greenish marbling UN produced by yellow-and-black scales closely approximated. Many have seasonal brood forms. Some are di- or polymorphic (two or more forms, which may be sex determined). California members of this family are nonmigratory; some tropical species migrate in swarms numbering thousands. *Early stages*: *egg* flask shaped, taller than wide; *larva* slender, cylindrical, smooth or granulated; *pupa* with pointed head and large wing cases, attached head up and suspended by a silken girdle. *Larval food plants*: of whites, mostly members of the Mustard Family (Brassicaceae); of many sulfurs, members of the Pea Family (Fabaceae), such as alfalfa and clover.

Pine White   (*Neophasia menapia*) (Pl. 21a and Figure 28). White, about the size of the Cabbage Butterfly; two-thirds of FW costal margin black; UNHW veins often red margined in female. Our only member of the Pieridae associated with conifers. Found in pine forest as far south as the Tehachapi Mountains in Kern County, and as far north as British Columbia. In outer Coast Ranges of Sonoma and Mendocino counties is found the Coastal Pine White (*N. m. melanica*); the black submarginal band wider, the black cell-bar narrower. Flight period June–August; one brood. *Early stages*: *egg* laid in rows at base of pine needles, green, flask-shaped, a circle of beads at tip, fluted at sides; *larva* dark green, a white band on each side and down the back, cylindrical, two short anal tails: *pupa* slender, dark green, whitestriped. *Larval food plants*: pines, including Yellow Pine (*Pinus ponderosa*), Jeffrey Pine (*P.*

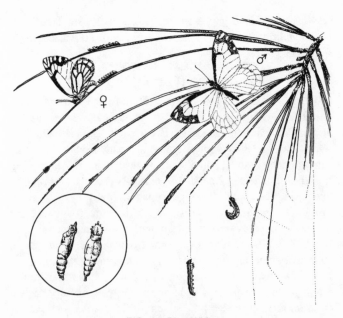

FIG. 28 Pine White

jeffreyi), and Lodgepole Pine (*P. murrayana*), also Douglas Fir (*Pseudotsuga menziesii*) (Pinaceae).

Becker's White (*Pontia beckerii*) (formerly *Pieris*) (Pl. 21b). Found in brushlands east and north of the Sierra Nevada and in the desert areas of southern California, from which it invades the coastal lowlands through low mountain passes. Black cell-bar wide, often white centered; veins of UNHW green edged. Flies all warm season; multiple broods. *Early stages*: *larva* green, with black tubercles and narrow orange bands; *pupa* gray or brown, with blunt front end; wing cases lighter. *Larval food plants*: various mustards (*Brassica* spp.), Hedge Mustard (*Sisymbrium officinale*), Tumble Mustard (*S. altissimum*), prince's plume (*Stanleya* spp.), and others (Brassicaceae); Bladder Pod (*Isomeris arborea*) (Capparaceae).

California White (*Pontia sisymbrii*) (formerly *Pieris*) (Pl. 21c). In spite of its name, uncommon in many parts of

California. A small white, the wing veins narrowly dark both UP and UN. Coast Range hills (uncommon); mountains up to 9,000 ft.; deserts east of Sierra Nevada; Mojave Desert and dry coastal hills of southern California; usually absent from the Great Central Valley and farm lands. Flies early (March–April) in the deserts, later (May–July) in the mountains; one brood. *Early stages*: *larva* yellow with narrow dark bands; *pupa* dark brown, surface granulated. *Larval food plants*: rock cress (*Arabis* spp.); rock cabbage (*Caulanthus* spp.), jewel flower (*Streptanthus* spp.), Hedge Mustard (*Sisymbrium officinale*), from whence its specific name, and other mustards (Brassicaceae).

Common White, Checkered White   (*Pontia protodice*) (formerly *Pieris*) (Pl. 21d). Formerly in every backyard and farm lot two white butterflies were found most of the year. One was the introduced Cabbage Butterfly, the other the native Common or Checkered White. With increasing urbanization and loss of open spaces, these butterflies are no longer common in our cities, but may still be found in our shrinking rural areas. Glossy white, the pointed FW with dark markings, the HW usually all white in the male. Female duller with brown checkered markings. Spring brood smaller, more lightly marked, flies very early in the year. Very common in the coastal and desert regions of southern California, along the central California coast, in the Central Valley; less common in northern California and east of the Sierra Nevada, where it may be replaced by the Western White. *Early stages*: *egg* spindle shaped, green or yellow; *larva* slender, light to deep green, with four yellowish stripes and many small black dots; *pupa* yellowish to bluish gray, with points of pink and yellow and many small black dots. *Larval food plants*: many members of the Mustard Family, including mustards (*Brassica* spp.), Hedge Mustard (*Sisymbrium officinale*); in the deserts, Desert Alyssum (*Lepidium fremontii*), and many others (all Brassicaceae): also Yellow Bee Plant (*Cleome lutea*) (Capparaceae).

Western White   (*Pontia occidentalis*) (formerly *Pieris*) (Pl. 21e). Closely resembling the Common White, the Western

White male has a dark spot in cell Cu$_2$ of FW (this spot lacking in male Common White), and both sexes are more heavily marked. The brownish markings found in female Common White are lacking in Western White; the veins of UNHW of Western White are outlined in greenish gray. Absent from the southern part of the state; more common northerly. A well-known "hilltopper." Flight period May–September; two or three broods annually. The Calyce White, form *"calyce"* (Pl. 21f), smaller, more heavily marked, is found at high elevations in the Sierra Nevada. It was once considered a subspecies. *Early stages*: *egg* and *pupa* similar to Common White; *larva* dull green, with narrow light and dark bands. *Larval food plants*: similar to those of the Common White.

Mustard White   (*Artogeia napi*) (formerly *Pieris*). Found in cool parts of both the Old World and the New World, the Mustard White frequents streamsides and woodlands. It is one of the earliest butterflies to appear in the spring. Adults variable, both seasonally and geographically. It may be recognized in any form by these markings: UNHW costa yellow at base; no dark FW cell bar; and no dark HW costal bar. California has three subspecies: (1) Veined White (*A. napi venosa*) (Pl. 21g), dark veined; Coast Ranges of central California, February–May. (2) Small-veined White (*A. n. microstriata*) (Pl. 22a), the veins narrowly gray; inner Coast Ranges and Sierra Nevada foothills, March–May. (3) Margined White (*A. n. marginalis*) (Pl. 21h), less heavily marked; northern California, March–May; its second brood, June–July, is the almost entirely white Pallid White (form *"pallida"*). *Early stages*: *egg* yellow green, pear shaped, and ribbed vertically; *larva* green with dark dorsal stripe, to bright green with three longitudinal stripes; *pupa* pale green or whitish, finely speckled with black. *Larval food plants*: most often Milkmaids (*Dentaria californica*), but also rock cress (*Arabis* spp.), winter cress (*Barbarea*), and very often, Water Cress (*Rorippa nasturtium-aquaticum*) (Brassicaceae).

Cabbage Butterfly   (*Artogeia rapae*) (formerly *Pieris*) (Pl. 22b). Accidentally introduced into North America, the Cabbage Butterfly has spread to all parts of the country. Also called the Imported Cabbage Worm, it is one of the principal pests of cabbage and other cole vegetables. It may be recognized by the chalky white color, the dark FW tips, the lack of a FW cell bar, and the dark spot at the outer two-thirds of HW costa. The lack of FW cell bar is shared with the Mustard White, but the costal spot is unique. Many brooded, reproducing all year except in very cold weather. *Early stages*: *egg* pale green, pear shaped, the small end fastened to the plant; *larva* velvety, bright green, usually with a dark line down the back and yellow dots on the sides, and with many small black dots; *pupa* either bright green or pale brown. *Larval food plants*: most members of the Mustard family (Brassicaceae), including all cultivated cole vegetables; also Garden Nasturtium (*Tropaeolum majus*) (Tropaeolaceae).

Alfalfa Butterfly, Common Sulfur, Orange Sulfur   (*Colias eurytheme*) (Pl. 11a). Because it found cultivated alfalfa as suitable for larval food as the native plants on which it originally fed, this native sulfur has become a major pest of this important crop. It is extremely variable in color. Males are yellow or orange with a solid black border. Females are yellow, orange, or white (form "*alba*"), with a broken border. Spring forms (form "*ariadne*") have only a faint orange tint. Summer forms (form "*amphidusa*") may be bright orange, often with a lavender sheen in the male. Very late and very early individuals are quite small, often dull yellow, and are form "*autumnalis*." Flight period all year in warm climates, February–November in valleys of California, later at higher elevations; several broods a year. *Early stages*: *egg* fusiform (cylindrical, tapering evenly to each end), red or orange, darkening before hatching; *larva* light to dark green, with pale stripes or red lines along the sides; *pupa* robust, green with black points and short yellow lines. *Larval food plants*: many members of the Pea Family, including Alfalfa (*Medicago sativa*), vetch (*Vicia* spp.), Deer Weed (*Lotus scoparius*), clover (*Trifolium* spp.), rattleweed and locoweed (*Astragalus* spp.), and others (Fabaceae).

Clouded Sulfur, Yellow Sulfur   (*Colias philodice eriphyle*) (Pl. 11b). Similar in size and markings to the Alfalfa Butterfly, but lemon yellow, without orange tint. HW cell spot usually yellow rather than orange. White females may be separated by the yellow HW cell spot and by the slightly wider black FW border. Often found flying with the Alfalfa Butterfly in eastern California, where their ranges overlap. In northern California, found mostly east of the Sierra-Cascade divide, the most notable exception being Scott Valley, Trinity County. In southern California, found near Holtville, Imperial County. Often abundant in cultivated lands. Flight period long; several broods. *Early stages* and *Larval food plants*: similar to those of the Alfalfa Butterfly; clover (*Trifolium* spp.) (Fabaceae) said to be preferred over Alfalfa.

Black and Gold Sulfur, Golden Sulfur   (*Colias occidentalis chrysomelas*) (Figure 29). A large and beautiful sulfur. The male is clear lemon yellow with wide, intense black borders. The female is paler yellow, the borders broken, faint, or nearly absent. This fine species has been found in the Coast Ranges from Sonoma County northward and in the Sierra Nevada–Cascade Range from Mariposa County north. It is local,

FIG. 29 Black and Gold Sulfur

usually uncommon to rare. Flight period May–June; one brood. *Early stages*: undescribed. *Larval food plants*: vetch (*Vicia* spp.) reported (in Washington); lupine (*Lupinus* spp.) and rattleweed (*Astragalus* spp.) suspected; closely associated with wild pea (*Lathyrus* spp.) in Sonoma County (Fabaceae).

Harford's Sulfur   (*Colias harfordii*) (Pl. 11c). The black-and-yellow sulfur of southern California, where it occurs in coastal ranges and valleys and the Transverse Ranges as far north as Santa Barbara and Kern counties. Distinctive spring forms occur at Santa Barbara and in San Diego County near Scissors Crossing. The name *barbara* was at one time applied to the first of these. Flight periods are March–May and June–August; two broods. Adults are attracted to blossoms of mint and thistle on dry hillsides. *Early stages*: *egg* yellow green, fluted, tapering; *larva* green with white-edged vermilion line along sides; *pupa* yellow green, the abdomen yellow banded and with a red line. *Larval food plant*: Douglas's Rattleweed (*Astragalus douglasii* var. *parishii*) (Fabaceae).

Edwards's Sulfur   (*Colias alexandra edwardsi*) (Pl. 11d). A large yellow sulfur, with narrow black border in the male, the border faint or lacking in female. A small black FW cell spot; HW cell spot pale or inconspicuous. UNHW dusted with greenish gray, cell spot small and pale. This handsome and fast-flying sulfur is found in northern California, mostly in the brushlands east of the Sierra–Cascade divide. Flight period June–August; one brood. *Early stages*: *larva* yellow green, a white lateral stripe enclosing orange dashes. *Larval food plants*: rattleweed (*Astragalus* spp.), False Lupine (*Thermopsis macrophylla*), White Clover (*Trifolium repens*), Alfalfa (*Medicago sativa*), and other legumes (Fabaceae).

Behr's Sulfur   (*Colias behrii*) (Pl. 11e). Though common in its chosen habitat, the alpine meadow, this unusual-looking little sulfur is quite restricted in its total range, being found only at high elevations in the Sierra Nevada, from Tulare and Inyo counties north to Tuolumne County. It is common in the vicinity of Tioga Pass. The male is greenish, the female lighter; albinic females are lighter still. Flight period July–August; one brood. *Early stages*: *larva* green, a dark-edged pink dorsal line, a light-edged salmon-colored lateral line, and dark spots on each segment. *Larval food plants*: Dwarf Bilberry (*Vaccinium caespitosum*) (Ericaceae); record of Alpine Gentian (*Gentiana newberryi*) (Gentianaceae) probably needs verification.

California Dog-face, Flying Pansy (*Zerene eurydice*) (Pl. 12a). The State Butterfly of California, the Dog-face is considered by many to be our most beautiful butterfly. The male is a rich yellow with violet reflections, and the black borders outline a fanciful dog's head, complete with eye, on each forewing. The female is normally entirely yellow with one round black spot on each forewing. The California Dog-face is widely distributed at low to moderate elevations, but is quite local, being common in relatively few places. Unlike the Southern Dog-face, it is never found in desert regions. Flies in April–June and July–September; double brooded. The second brood is especially notable in the San Bernardino and San Jacinto mountains of southern California. Males with dark-bordered HW (form "*bernardino*") are frequently encountered. Occasionally females with the dog's-head outlined (form "*amorphae*") are seen. A strong flier, it is usually difficult to catch, except when it stops at flowers, its food plant, or fresh horse manure, for which it has an incurable addiction. *Early stages*: *egg* fusiform, green; *larva* dull green with black dots, a pale lateral line tinted with orange, and above this a dark mark on each segment; *pupa* curved, with pointed head and large wing cases; bright green. *Larval food plant*: False Indigo (*Amorpha californica*) (Fabaceae).

Southern Dog-face (*Zerene cesonia*) (Pl. 12b). Differs from the California Dog-face in having both sexes black bordered and lacking the purplish sheen in the male. The Southern Dog-face occurs in the desert ranges of San Diego, Imperial, Riverside, and San Bernardino counties and adjacent areas west of the mountains. Flight period April–June and September–October; two broods. *Early stages*: *egg* yellow green, spindle shaped; *larva* green with black, hair-studded tubercles, often striped yellow and black; *pupa* overwinters. *Larval food plants*: Western False Indigo (*Amorpha fruticosa* var. *occidentalis*), indigo bush (*Dalea* spp.), clover (*Trifolium* spp.), no doubt others (Fabaceae).

Cloudless Sulfur   (*Phoebis sennae marcellina*) (Pl. 11f). This striking insect, once abundant in coastal southern California, has been absent there for several decades. During this period it was restricted to the desert portions of the state, from Palm Springs to the Providence Mountains and sparingly north to central California. It was also once known to breed in San Jose. Recently (in 1983) it reappeared in the coastal portions of its range in limited numbers. Males are a brilliant sulfur yellow above, without dark markings. Females may be either orange yellow or dull white, with a black cell-spot and dark borders. Flight period April–May in the desert; late summer in coastal regions; several broods a year. *Early stages*: *egg* orange yellow, spindle shaped, with vertical ribs; *larva* yellow green, a yellow stripe on each side, and segmental dark spots; *pupa* curved, with a long, pointed head and very large wing cases. *Larval food plants*: senna (*Cassia* spp.), including native Armed Cassia (*Cassia armata*), and cultivated species (Fabaceae).

Large Orange Sulfur   (*Phoebis agarithe*) (Pl. 11g). Differs from Cloudless Sulfur in having male orange instead of yellow. The female has a lighter ground color and is quite heavily marked. Found in eastern San Bernardino and Imperial counties, which it invades from Arizona. May be seen in desert mountain ranges such as the Providence Mountains, or in the Imperial Valley, as at Bard. *Early stages*: undescribed for the U.S. *Larval food plants*: black-bead (*Pithecellobium* spp.), senna (*Cassia* spp.) (Fabaceae).

Mexican Yellow   (*Eurema mexicana*) (Pl. 12k). An invader from Mexico and Arizona, this delicately shaded yellow with the short triangular tails is found in the southern counties—San Diego, Imperial, Orange, Riverside, and San Bernardino—and east to Texas. Common in the canyons on the desert side of mountain ranges, it occasionally penetrates low passes like Santa Ana Canyon to reach the coast. Flight period May–November; multiple brooded. *Early stages*: unknown. *Larval food plant*: reported to be senna (*Cassia* spp.) (Fabaceae).

Nicippe Yellow, Sleepy Orange   (*Eurema nicippe*) (Pl. 12h). Formerly common in populated areas, this deep orange species with the heavy black margins is found throughout southern California in coastal lowlands as far north as Santa Barbara County, and in deserts, where large fall broods follow summer rains. Reasons for its disappearance from city parks and gardens may be increased air pollution, or use of insecticides. Flight period April–November; two–three broods. *Early stages*: *egg* spindle shaped, ribbed, greenish yellow; *larva* green, yellow striped; *pupa* olive green. *Larval food plants*: senna (*Cassia* spp.) (Fabaceae); native species are used in the desert, introduced species in cities and towns.

Dainty Sulfur   (*Nathalis iole*) (Pl. 12l). A tiny sulfur, about the size of a blue. FW apex and inner margin, and HW costa and vein tips black. Male ground color pale yellow, a sex spot on HW costa. Female more buffy. A species of extensive range, often uncommon or overlooked. In northern California, known only from east of the mountains, and on the east slope of the Sierra Nevada to tree line. In southern California, desert areas of Riverside, San Bernardino, and Imperial counties, and coastal areas of Orange and San Diego counties. Flight period March–October; multiple brooded. It turns up at unexpected times and places, such as at 9,000 ft. on Mt. San Jacinto in March! *Early stages*: *larva* dark green with purple dorsal stripe, and yellow-and-black lateral stripes; *pupa* green, without frontal projections. *Larval food plants*: a wide variety of unrelated plants; in northern California thought to be filaree or storksbill (*Erodium* spp.) (Geraniaceae); in southern California Beggar-tick or Bur Marigold (*Bidens pilosa*), also garden marigold (Asteraceae); no less than six other plants reported from other parts of its range.

Felder's Orange-tip   (*Anthocharis cethura*). No butterfly is more clearly identified with arid regions of the Mojave and Colorado deserts than Felder's Orange-tip, nor is more eagerly sought after during the month of April in which it flies. Found from San Diego and Imperial counties on the south to Kern and Inyo counties on the north, it inhabits dry washes, such as the San Felipe or Morongo, and desert buttes, such as Alpine and Lovejoy, where it often hilltops. Several subspecies occur in its range: (1) Felder's Orange-tip (*A. c. cethura*) (Pl. 12c), found at Little Rock, near Palmdale, and adjacent areas. (2) Morrison's Orange-tip (*A. c. morrisoni*), UNHW green mottling heavy, female lacking orange tip, northern part of range, as lower Kern River area. (3) Caliente Orange-tip (*A. c. caliente*), a yellow form, occasional in all populations but increasingly to the east. (4) Desert Orange-tip (*A. c. deserti*), orange tip reduced or absent, southernmost part of the range. (5) Catalina Orange-tip (*A. c. catalina*), the only completely segregated subspecies, found only on Santa Catalina Island. (6) Pima Orange-tip (*A. c. pima*) (Pl. 12e), deep yellow with bright orange red tips, far eastern California and Arizona. Some workers consider all except (1), (5), and (6) to represent forms rather than subspecies. Flight period March–April; one brood. *Early stages*: *egg* orange, spindle shaped, ribbed; *larva* cross-banded orange and green, with rows of black dots on the cross-bands; *pupa* brown to gray with fine dark striations. *Larval food plants*: Long-beaked Streptanthella (*Streptanthella longirostris*), Western Tansy Mustard (*Descurainia pinnata*), and Desert Candle (*Caulanthus inflatus*), perhaps others (Brassicaceae).

Sara Orange-tip   (*Anthocharis sara*) (Pls. 12d, 12g). The glimpse of an orange-tip dashing along in the sunshine is a sure sign of spring. The white ground color, the reddish orange FW tips, and the mossy UN marbling are distinctive. The spring brood, long called Reakirt's Orange-tip (form "*reakirtii*") (Pls. 1e, 1f, 12f, 12j) flies in February–April. The second brood (*sara*) flies in May–June. It is believed the chrysalids from the second brood overwinter and the spring "*reakirtii*" adults emerge from these. California has four subspecies: (1) Sara Orange-tip (*A. s. sara*), found in much of California, except where replaced by other subspecies. (2) Stella Orange-tip (*A. s. stella*), both males and females yellowish, high elevations in the Sierra Nevada. (3) Thoosa Orange-tip (*A. s. thoosa*), white, the dark marking heavier, sometimes slightly brownish, east of Sierra Nevada and Mojave Desert ranges. (4) Gunder's Orange-tip (*A. s. gunderi*), male heavily black barred, female dark apically, both sexes HW greenish yellow, Santa Cruz and Santa Catalina islands. Yellow forms occur now and then in the normally white populations. An odd-looking orange-tip named "*dammersi*," of which only the type and two other specimens have ever been found, has been considered by some to be a hybrid between the Sara Orange-tip and Grinnell's Marble, or an aberration of the Sara Orange-tip. *Early stages*: egg taller than wide, with a small base, silvery green; *larva* green, a light stripe down the back; *pupa* curved and pointed at the head end, either pale brown or green. *Larval food plants*: many members of the Mustard Family (Brassicaceae), including Hedge Mustard (*Sisymbrium officinale*), rock cress (*Arabis* spp.), Western Tansy Mustard (*Descurainia pinnata*), winter cress (*Barbarea* spp.), Lace Pod (*Thysanocarpus curvipes*), Black Mustard (*Brassica nigra*), Field Mustard (*Brassica campestris*), Charlock (*Brassica kaber*), and others.

Boisduval's Marble   (*Falcapica lanceolata*) (formerly *Anthocharis*). No orange tips. FW falcate (pointed); UN very finely marbled with gray or dull brown. Found along canyon walls where streams come out of the mountains, or along semi-desert arroyos. An uncommon and elegant species, of rapid flight. Two subspecies occur in California: (1) Boisduval's Marble (*F. l. lanceolata*) (Pl. 22c), found around rocky cliffs and outcrops at low elevations over much of northern California, April–May, up to 8,000 ft. in southern Sierra Nevada in June–July. (2) Grinnell's Marble (*F. l. australis*) (Pl. 22d), found in wooded canyons in lower mountain areas south of

Kern River, Kern County, and also along western edge of Mojave and Colorado deserts in March to May; one brood. *Early stages*: *larva* green with white flecks and lateral line; *pupa* gray brown, the palpal case enclosing the mouthparts strongly curved. *Larval food plants*: rock cress (*Arabis* spp.), including Elegant Rock Cress (*S. sparsiflora* var. *arcuata*), Perennial Rock Cress (*A. perennans*), Tower Mustard (*A. glabra*), and others (Brassicaceae).

Edwards's Marble   (*Euchloe hyantis*). Usually uncommon, it flies in most of the same places and at the same time as Boisduval's Marble. Compared to the Large Marble, Edwards's Marble is smaller, more silky white, and has heavier green marbling UN. Three subspecies in California, all with a single brood: (1) Edwards's Marble (*E. h. hyantis*) (Pls. 12i, 22e), found over much of northern California, including the Sierra Nevada. Flight period mid-April to mid-July. *Larval food plants* rock cress (*Arabis* spp.), jewel flower (*Streptanthus* spp.), hedge mustard (*Sisymbrium* spp.), and others (Brassicaceae). (2) Southern Marble (*E. h. lotta*) (Pl. 22f), from Red Rock Canyon and Palmdale on the Mojave Desert to Palm Springs on the Colorado Desert, east into the Rocky Mountains. Flight period March–April. *Larval food plants* as above plus rock cabbage (*Caulanthus* spp.), tansy mustard (*Descurainia* spp.), and Desert Alyssum (*Lepidium fremontii*) (Brassicaceae). (3) Martin's Marble (*E. h. andrewsi*) (Pl. 22g), San Bernardino Mountains near Lake Arrowhead and Big Bear Lake, 5,000–6,000 ft. Flight period late May–early July. *Larval food plants* Mountain Tansy Mustard (*Descurainia richardsonii*), Holboel's Rock Cress (*Arabis holboelii* var. *pinetorum*), and San Bernardino Jewel Flower (*Streptanthus bernardinus*) (Brassicaceae). *Early stages* (of *E. h. lotta*): *egg* lemon yellow; *larva* green with purplish bristle-tipped dots, white lateral line, and purple middorsal stripe; *pupa* light straw to dark brown. Only the seed pods of the plant are eaten by the larvae.

Large Marble   (*Euchloe ausonides*) (Pl. 22h). The commonest marble at low elevations in central and northern California. Ground color often tinged with yellow; some females are quite yellowish. UNHW without pearly luster, the marbling rather wide spaced. Found in meadows, fields, farm lands, vacant lots, and along streamsides. Common in the lower valleys; scarcer at high elevations. Flight period March–June; usually two broods. *Early stages*: *egg* blue green changing to light orange, long, ribbed vertically; *larva* dark green with dorsal and lateral yellow green stripes; *pupa* light brown, the wing covers and five abdominal lines darker. *Larval food plants*: rock cress (*Arabis* spp.), hedge mustard (*Sisymbrium* spp.), Winter Cress (*Barbarea orthoceras*), tansy mustard (*Descurainia* spp.), wall flower (*Erysimum* spp.), Black Mustard (*Brassica nigra*), Field Mustard (*B. campestris*), and others (Brassicaceae).

The internationally famous pianist, Walter Gieseking, once kept an Oakland audience waiting for two hours in a spring afternoon while he indulged his favorite pastime, butterfly collecting, in the Berkeley hills. To those "in the know" it was apparent that he was looking for, and probably found, the Large Marble.

Snout Butterflies (Family Libytheidae)

Palpi prolonged beyond the head to form a "snout"; front legs of male, but not of female, reduced; larva slender, cylindrical, without spines. A small and ancient family, comprising but ten species worldwide, one of which occurs in our area.

Snout Butterfly   (*Libytheana bachmanii larvata*) (Pl. 3d). Size 1⅜–1⅞ in. (34–47 mm). An orange-and-brown butterfly with square-cut wing tips and conspicuous white spots UPFW, plus a pair of elongated palpi that extend almost as far forward as the antennae. Occurring as a stray in southern California, the Snout Butterfly is found in areas bordering the Mojave and Colorado deserts from Inyo County south, and Los Angeles County east, and occasionally to the coast (Orange County), to which it has low-elevation access through San Gorgonio Pass and Santa Ana Canyon. A resident population is established near Banning, in Riverside County, where native hackberry grows. Flight period (in California) September–November; probably two broods in the west; multiple broods elsewhere. *Early stages*: *egg* pale green, tapered at both ends; *larva* dark green with yellow dorsal and lateral lines; *pupa* green with yellow lines and spots. *Larval food plant*: Netleaf Hackberry (*Celtis reticulata*) (Ulmaceae).

Metalmarks (Family Riodinidae)

Small butterflies, size ¾–1¼ in. (19–31 mm), with metallic markings, relatively long antennae, and a spur vein in HW. Closely related to the Lycaenidae, with which they share reduced forelegs of male, notched eyes around antennal bases, and flattened larvae. First segment of reduced male foreleg extends beyond second; wings held horizontally when at rest. Largely tropical; only one of California's four species extends into northern California.

Mormon Metalmark   (*Apodemia mormo*). A small black and orange-brown butterfly checkered with white. Size ⅞–1¼ in. (22–31 mm). Found in dry waste places; often overlooked. In northern California, single brooded, adults July–September; in southern California, double brooded, adults March–May and August–October. California has seven of nine recognized subspecies: (1) Mormon Metalmark (*A. m. mormo*) (Pl. 23g), found in northern California and eastern San Bernardino and Riverside counties in dry and rocky habitats. (2) Lange's Metalmark (*A. m. langei*) (Pls. 13b, 23h), orange brown of UPHW wide and bright, white spots small; restricted to the vicinity of Antioch, Contra Costa County, an endangered population. (3) Tuolumne Metalmark (*A. m. tuolumnensis*), UPFW broadly orange, white spots reduced; Pate Valley, Grand Canyon of the Tuolumne; very scarce. (4) Cythera Metalmark (*A. m. cythera*) (Pl. 23k), UPFW and UPHW broadly pale orange, UNHW pale; eastern slope of Sierra Nevada south to Walker Pass and Piute Mountain, Kern County. (5) Behr's Metalmark (*A. m. virgulti*) (Pls. 1m, 1n, 13a, 23i), UPFW and UPHW red orange; coastal ranges of southern California. (6) Desert Metalmark (*A. m. deserti*) (Pl. 23j), light markings

UPFW and UPHW reduced, pale orange; UNHW light; Morongo Wash and Palm Springs south to Anza-Borrego Desert. (7) Whitish Metalmark (*A. m.* nr. *dialeuca*) (Pl. 231), orange suppressed, white spots conspicuous; high mountains of San Bernardino County. *Early stages*: *egg* pink; *larva* violet with four rows of tufted spines with black bases, and longer ochre spines; overwinters; *pupa* mottled brown, its surface hairy. *Larval food plants*: wild buckwheats (*Eriogonum* spp.), including California Buckwheat (*E. fasciculatum*), Wright's Buckwheat (*E. wrightii*), Naked Buckwheat (*Eriogonum nudum*), Broadleaved Buckwheat (*E. latifolium*), Desert Trumpet (*E. inflatum*), and no doubt many others (Polygoniaceae).

Palmer's Metalmark   (*Apodemia palmeri marginalis*) (Pls. 13c, 23m). A true child of the desert, this small but colorful metalmark is found sparingly in the Coachella and Imperial valleys of eastern Riverside, San Diego, and Imperial counties, where it frequents oases such as Thousand Palms and Indian Wells, and sometimes roadside food plants. Size ¾–1 in. (19–25 mm). Flight period April–November; two or three broods, depending on rainfall. *Early stages*: *egg* green, hemispherical; *larva* bluish green, with pale yellow dorsal and lateral stripes and six tufts of white hair on thorax; *pupa* green with yellow wing cases. *Larval food plants*: Honey Mesquite (*Prosopis glandulosa* var. *torreyana*), Screwbean Mesquite (*P. pubescens*) (Fabaceae).

Fatal Metalmark   (*Calephelis nemesis*). A small brown metalmark, the UPFW and UPHW crossed by a wide, indistinct dark brown band. California has two subspecies: (1) Dusky Metalmark (*C. n. californica*) (Pl. 13e), size ¾–1 in. (19–25 mm), coastal mountains in the chaparral zone of Los Angeles, Orange, Riverside, and San Diego counties. Flight period February–October; three broods. (2) Dammers's Metalmark (*C. n. dammersi*), lighter and smaller, Colorado River bottom near Blythe, Riverside County. Flight period July; one brood (?), little known. *Early stages*: *egg* echinoid (sea urchin shaped); *larva* gray with long whitish or brownish hairs in two rows down the back and one row on each side; *pupa* yellowish with brown dots, suspended by tip of abdomen and supported by a thread around thorax. *Larval food plants*: Mule Fat (*Baccharis glutinosa*) for interior populations, California Encelia (*Encelia californica*) for coastal populations (Asteraceae).

Wright's Metalmark (*Calephelis wrighti*) (Pl. 13d). Larger than the preceding species, and with a distinctly reddish cast to the brown UP. Wright's Metalmark flies in the Colorado Desert of eastern Riverside and San Diego counties east into Arizona and south into Mexico. It also occurs in a few coastal areas in San Diego County. Size ⅞–1 in. (22–25 mm). Flight period March–October; three brooded in years of summer and fall rains. *Early stages*: *egg* echinoid; *larva* grayish white, nodular (lumpy), with four rows of long white hairs, spiracles (breathing openings) yellow; *pupa* greenish or brownish, tuberculate (bumpy), and short-hairy. *Larval food plant*: Sweet Bush (*Bebbia juncea*) (Asteraceae), of which the larva eats only the greenish outer layer of the stem.

Hairstreaks, Coppers, and Blues (Family Lycaenidae)

Small butterflies, size ½–1¾ in. (13–44 mm), the antennae arising from notches at the upper corner of each eye. Radius (main stem vein) of forewing has three or four branches; front legs of male reduced, those of female full sized. *Egg* biscuit shaped (said to be turban shaped); *larva* has head partly concealed in thorax, body short, robust, and flattened below (said to be slug shaped), usually covered with very short velvety hair; *pupa* short and rounded, with a deep depression between thorax and abdomen, appressed closely to the surface where it is attached by tip of abdomen and by a silk strand around the middle.

The Lycaenidae divide nicely into three groups. Hairstreaks (subfamilies Theclinae and Eumaeinae) have pointed FW and usually have HW lobed at anal end (tornus); they may have short, hairlike tails. Males often have a stigma (sex mark) on FW. Coppers (subfamily Lycaeninae) are larger than most lycaenids; UN is usually covered with many small spots. Sexual dimorphism is great; the females are duller and more spotted than the males. Blues (subfamily Polyommatinae) (formerly Plebejinae) average small; differences between male and female are often extreme; however, the spot pattern UN on any given species is the same in both sexes. Lycaenid larvae have glands that secrete honeydew and are commonly tended by ants in exchange for this delicacy.

Subfamily Theclinae

FW radius four-branched as in Lycaeninae; eyes hairy; palpi slanting, third segment long, scaled. Male without FW stigma. HW tornus rounded, with short, slender tails. One species in our area.

Boisduval's Hairstreak, Canyon Oak Hairstreak   (*Habrodais grunus*). UP brown; UN yellow brown; UNHW has a row of narrow, dark crescents, the last two green. Size about 1 in. (25 mm). Flight period late June–August; one brood. Adults crepuscular, active in early morning and evening hours; usually rest in shade the rest of the day. Three subspecies: (1) Boisduval's Hairstreak (*H. g. grunus*); Sierra Nevada and Transverse ranges of southern California. (2) Lorquin's Hairstreak (*H. g. lorquini*) (Pl. 13g), darker; Coast Ranges of Central California. (3) Herr's Hairstreak (*H. g. herri*), UN very light; northern California. *Early stages*: *egg* deposited on oak twig; *larva* bluish green with a yellow median stripe, hairy; *pupa* also bluish green and hairy, with brown dots. *Larval food plants*: various oaks, especially Canyon Live Oak (*Quercus chrysolepis*) (Fagaceae).

Subfamily Eumaeinae

FW radius three-branched; eyes hairy; palpi with second segment often not hairy, more smoothly scaled; third segment slender, often quite long. Male often with FW stigma. HW tornus lobed, often with one or two slender tails. Includes the rest of the hairstreaks. Coloration often dull gray or brown UP, with bright markings UNHW.

Sarita Hairstreak   (*Chlorostrymon simaethis sarita*) (Pl. 13h). UP grayish brown, overlaid with brilliant purple; UN bright green with an irregular white line and purplish area at base of tails. Size ⅞–1⅛ in. (22–28 mm). A resident of southern Texas, Mexico, and Baja California. Garth encountered it at San Gabriel Bay, Espiritu Santo Island, in the Gulf of California. This delightful little hairstreak occurs as a straggler in southern California, in San Diego County, in October. *Early stages*: undescribed. *Larval food plants*: Balloon Vine (*Cardiospermum halicacabum*) and Heartseed (*H. cornidum*), of which the larva eats the developing seeds (Sapindaceae).

Coral Hairstreak   (*Harkenclenus titus immaculosus*) (Pl. 13f). Dark brown UP and UN, a submarginal orange band in female, outer edge UNHW with a row of coral red spots; no tails on HW. Size 1–1¼ in. (25–31 mm). Flight period July–August; one brood. Occurs uncommonly at low and moderate elevations in the Sierra Nevada. *Early stages* (of *H. t. titus*; apparently not known for *immaculosus*): *egg* green, covered with raised lines; *larva* yellowish green with a dark green dorsal line and pink patches, hairy; *pupa* brown, with darker brown dots. *Larval food plants*: wild cherry, wild plum (*Prunus* spp.) (Rosaceae).

Sooty Gossamer-wing   (*Satyrium fuliginosum*) (Pl. 13i). UP dark gray to brownish gray, UN slightly lighter; light spots few or lacking; no tails. Size 1–1¼ in. (25–31 mm). Flight period late June–early August; one brood. Found mostly east of the Sierra-Cascade crest, but also in Siskiyou and Trinity counties, often in open brushland, from Mono Lake north. Locally common but very inconspicuous, and distribution apparently spotty. *Early stages*: unrecorded. *Larval food plants*: lupine (*Lupinus* spp.) (Fabaceae).

Behr's Hairstreak   (*Satyrium behrii*) (formerly *Callipsyche*) (Pl. 13j). UP yellow brown with dark brown borders; UN with irregular small dark spots, each with a white border on its outer edge. Size 1–1¼ in. (25–32 mm). Flight period June–August; one brood. Inhabits mixed brushland, pinyon-juniper woodland, mostly east of the Sierra-Cascade divide, with scattered colonies on the north slopes of Tejon, Tehachapi, and San Bernardino mountains in southern California. *Early stages*: *larva* green with yellowish hairs and lines of white, dark green, and yellow; *pupa* tan blotched with brown. *Larval food plants*: in northern California, Antelope Brush (*Purshia tridentata*); in southern California, Waxy Bitterbrush (*P. glandulosa*) (both Rosaceae).

The remaining species of *Satyrium* were formerly in *Strymon*.

Gold-hunter's Hairstreak   (*Satyrium auretorum*). Deep brown (male) to golden brown (female); small blue spot UNHW at anal angle. Male FW pointed; HW short tailed. Size 1 – 1¼ in. (25 – 31 mm). Flight period May – early July, depending on locality; one brood. Foothills and moderate elevations, Coast Ranges and west slope of Sierra Nevada; also southern California. Local, usually uncommon to scarce. Two subspecies: (1) Gold-hunter's Hairstreak (*S. a. auretorum*), as described above (Pl. 13l). (2) Nut-brown Hairstreak (*S. a. spadix*) (Pl. 13m), from the Greenhorn Mountains south to the Laguna Mountains. At times abundant, as at Crystal Springs, San Gabriel Mountains, or at Cajon Pass, San Bernardino Mountains. *Early stages*: *egg* mauve with green spicules; *larva* green to orange with chestnut hairs; *pupa* buff with brown spots. *Larval food plants*: oaks, especially Blue Oak (*Quercus douglasii*) and Scrub Oak (*Q. dumosa*), also Interior Live Oak (*Q. wislizenii*), probably others (Fagaceae).

Gray Hairstreak   (*Satyrium tetra*) (Pl. 13k). UP brownish gray; UN lighter gray, either unmarked, or crossed by a vague light line; short tails. Size 1¼ – 1½ in. (31 – 38 mm). Flight period June – July; one brood. Found in chaparral of inner Coast Ranges and Sierra Nevada foothills; also on coastal side of Transverse Ranges of southern California. Formerly known as *Strymon adenostomatis*. *Early stages*: *larva* light green with segmentally arranged bluish bars and orange hairs; *pupa* brown with blackish blotches. *Larval food plant*: Hard Tack, Mountain Mahogany (*Cercocarpus betuloides*); long thought to be Chamise (*Adenostoma fasciculatum*) (both Rosaceae).

Hedge-row Hairstreak   (*Satyrium saepium*). UP coppery brown, our only hairstreak so colored. HW tailed. Size 1 – 1⅛ in. (25–28 mm). Flight period late April–July, depending on elevation; one brood. Found in chaparral and at forest edges. Flies almost everywhere except at highest elevations. California has four weakly differentiated subspecies, which are regarded as forms by some workers: (1) Hedge-row Hairstreak (*S. s. saepium*), color as above; northern California, including Sierra Nevada. (2) Tawny Hairstreak (*S. s. fulvescens*), lighter both UP and UN; drier parts of southern Sierra Nevada. (3) Purplish Hairstreak (*S. s. chlorophora*) (Pl. 13n), richer and deeper, almost purplish brown, UN also darker; San Diego County, also Santa Cruz Island. (4) Bronzed Hairstreak (*S. s. chalcis*), UN postmedian band of spots reduced, inner two-thirds of wings darker, outer one-third lighter; at least Santa Cruz, San Mateo, Santa Clara, and Alameda counties. *Early stages*: *egg* grayish green; *larva* green with yellowish chevron markings; *pupa* brown with black dots. *Larval food plants*: California lilac (*Ceanothus* spp.), especially Buck Brush (*C. cuneatus*) (Rhamnaceae).

Sylvan Hairstreak, Woodland Hairstreak   (*Satyrium sylvinum*). UP light olive brown; UN light gray, crossed by a row of small black spots; orange and dull blue markings near anal angle of HW; tails present. Size 1¼–1⅜ in. (31–34 mm). Flight period May–August; one brood. Frequents streamsides, willow thickets; found in both valleys and mountains, except at highest elevations. Two subspecies in California: (1) Sylvan Hairstreak (*S. s. sylvinum*) (Pl. 13p), as noted above. (2) Desert Hairstreak (*S. s. desertorum*), coloration lighter, size often larger; eastern slope of Tehachapi Mountains north through Kern, Inyo, and Mono counties. *Early stages*: *egg* echinoid, soiled white tinged with olive green, laid singly or in groups on willow twigs; *larva* light green with dorsal white stripe and white diagonal bars on each segment; *pupa* greenish brown spotted with darker green and brown. *Larval food plants*: various species of willows (*Salix* spp.) (Salicaceae).

Dryope Hairstreak   (*Satyrium dryope*) (Pl. 13q). Lighter colored than the Woodland Hairstreak; FW shorter; tails lacking. Size 1⅛–1⅜ in. (28–34 mm). Flight period mid-May–early July; one brood. Frequents streamsides and willow thickets of inner Coast Ranges from San Francisco Bay south to the Tejon Range in Northwestern Los Angeles County; also found in Mono County east of the Sierra Nevada. Some authors consider the Dryope Hairstreak a subspecies of the Woodland Hairstreak, with a blend zone in the western San Gabriel Mountains. *Early stages*: *egg* light brown, laid singly on food plant; *larva* (northerly) green with light lateral chevrons, or (southerly) gray with brown dots and translucent spines, the head and tail ends green; *pupa* (north) greenish brown, (south) undescribed. *Larval food plants*: willows (*Salix* spp.) (Salicaceae).

California Hairstreak   (*Satyrium californicum*) (Pl. 13o). UP olive brown; yellow or orange markings on outer lower edge of FW; UN dark gray, a band of small round black spots across both wings. Anal angle of HW has red spots; tails present. Size 1⅛–1⅜ in. (28–34 mm). Flight period May–July; one brood. Found in open woodland, forest edges, and chaparral; frequents many kinds of flowers. Distribution general throughout state; in southern California mostly restricted to higher mountains. *Early stages*: *egg* light brown, laid in small groups (two–four) at base of leaflet of food plant; *larva* brown with large gray spots bordered by white chevrons; *pupa* brown, wing covers mottled with black. *Larval food plants*: oak (*Quercus* spp.) (Fagaceae); Buck Brush (*Ceanothus cuneatus*) (Rhamnaceae); in eastern Oregon, association with hardtack (*Cercocarpus* spp.) (Rosaceae) noted.

Leda Hairstreak   (*Ministrymon leda*) (Pl. 13r). UP gray with blue at base of both wings; UN light gray crossed by an irregular narrow red line; a red spot at base of tails. Size ¾–⅞ in. (19–22 mm). Flight period April–October; several broods. Garth took this species near the summit of Mt. Palomar, in early July, and at Sentenac Canyon, San Diego County, in mid-April. This delicately marked hairstreak occurs throughout the desert areas of San Diego County north into the Colorado Desert and eastern Mojave Desert. *Early stages*: *egg* green with a network of fine lines, deposited on mesquite flowers; *larva* green with yellow markings, covered with short brown hairs; *pupa* brown with black blotches. *Larval food plant*: Mesquite

(*Prosopis glandulosa* var. *torreyana*) (Fabaceae). The Ines Hairstreak (*M. l.* form "*ines*") (Pl. 13s), long regarded as a separate species, is now considered to be the fall brood of the Leda Hairstreak. It is quite different in appearance; UP is darker, the blue more intense, the wing bases darker than outer half of wings, red line and red at base of tails absent.

Skinner's Hairstreak   (*Mitoura loki*) (Pl. 14g). Resembles Juniper Hairstreak, but basal half of UNHW is partly brown, the median line lacks white edging, and the submarginal row of spots is continuous. Size 1–1¼ in. (25–31 mm). Flight period March–April and June–July; an occasional third brood in November. Western edges of the Colorado Desert from east end of San Bernardino Mountains into Baja California, with isolated populations in Riverside County (Gavilan Hills) and San Diego County (Jacumba Hot Springs). *Early stages*: *egg* light green with lighter raised portions; *larva* green, resembling that of Juniper Hairstreak; *pupa* dark brown, hairy. *Larval food plant*: California Juniper (*Juniperus californica*) (Cupressaceae); may be reared on garden cypress.

Thorne's Hairstreak   (*Mitoura thornei*) (not shown). Similar to Skinner's Hairstreak (*M. loki*), but darker; UNHW less contrasting, light areas less conspicuous, medial band less strongly toothed on outer edge; HW marginal area more strongly dusted with light glaucous blue. Size 1–1⅛ in. (25–28 mm). Flight period February–May and October; at least two broods. Known only from the vicinity of the TL, Little Cedar Canyon, Otay Mountain, San Diego County, where it flies in close association with its food plant. *Early stages*: *egg* echinoid, light green, laid singly on food plant; mature *larva* vivid green, each segment with a white crescent on each side of middorsal crest, a thin white line on each side above prolegs, body covered with fine brown hairs; *pupa* dark chestnut brown, covered with fine erect brown hairs. *Larval food plant*: Tecate Cypress (*Cupressus forbesi*) (Cupressaceae).

Siva Hairstreak   (*Mitoura siva*). A tailed hairstreak with green UN. Size 1–1¼ in. (25–31 mm). Flight period March–May and June–August, depending on locality; at least two broods. Four subspecies in California: (1) Siva Hairstreak (*M. s. siva*) (Pl. 14d), Providence, New York, Kingston, and Panamint Mountains; TL Fort Wingate, New Mexico. (2) Juniper Hairstreak (*M. s. juniperaria*) (Pl. 14e), UN with pearly overlay; western Mojave Desert from San Bernardino Mountains to Mt. Pinos, Ventura County, and Lake Isabella, Kern County. (3) Mansfield's Hairstreak (*M. s. mansfieldi*), UN grass green, pearly overlay lacking; Coast Ranges of San Luis Obispo, Santa Barbara, and Ventura counties. (4) Clench's Hairstreak (*M. s. chalcosiva*), UNHW brown, but with dark and light markings as in Siva Hairstreak (*M. s. siva*); Westgard Pass north to Carson City or beyond; Great Basin in California, Nevada, and Utah; TL Stansbury Mountains, Tooele County, Utah. All four fly in juniper woodland, usually close to host trees, or to flowers such as those of the Turpentine Bush (*Happlopappus linearifolius*). *Early stages*: *egg* with triangular network; *larva* green with yellow crescents dorsally, resembles twigs of juniper; *pupa* brown, hairy. *Larval food plants*: California Juniper (*Juniperus californica*); in eastern part of range, Utah Juniper (*Juniperus osteosperma*) (both Cupressaceae).

Barry's Hairstreak   (*Mitoura barryi*) (not shown). Closely resembles Nelson's Hairstreak (*M. nelsoni*), but with UNHW markings more distinct, median line irregular, as in Siva Hairstreak (*M. s. siva*), outer half of HW lighter than basal half. Size 1–1⅛ in. (25–28 mm). Found from Oregon south into northeastern California in juniper woodland. Flight period mid-April–late June; one brood. *Early stages*: unknown. *Larval food plant*: not known. In Oregon, Tilden found adults closely associated with Western Juniper (*Juniperus occidentalis*) (Cupressaceae).

Nelson's Hairstreak   (*Mitoura nelsoni*) (Pl. 14c). Male UP brown; rusty marks on lower corners of wings; females more rusty; UN lilac brown, an irregular light line (sometimes reduced) across both wings; gray spots with black caps on outer edge; tails present. Size ⅞–1⅛ in. (22–28 mm). Flight period May–July; one brood. Flies in coniferous forest over much of California. Males are territorial, actively "defending" perches at forest edges. *Early stages*: *egg* echinoid, green; *larva* bright green with darker raised areas bordered by yellowish crescents; *pupa* brown, hairy. *Larval food plants*: Incense Cedar (*Calocedrus decurrens*); in eastern California, perhaps Utah Juniper (*Juniperus osteosperma*) (both Cupressaceae).

Muir's Hairstreak   (*Mitoura muiri*) (Pl. 14f). Similar to Nelson's Hairstreak but darker, especially UN, and with the band much more irregular. Size ⅞–1⅛ in. (22–28 mm). Flight period May–June; one brood. Coastal California from Mendocino and Lake counties south to Santa Lucia Mountains, San Luis Obispo County. *Early stages*: apparently undescribed. *Larval food plant*: Sargent Cypress (*Cupressus sargentii*) (Cupressaceae). On Mt. Diablo, Contra Costa County, and south into Monterey County, there is a population of hairstreaks that resembles Muir's Hairstreak. It is uncommon. The most likely food plant there is California Juniper (*Juniperus californica*) (Cupressaceae).

Thicket Hairstreak   (*Mitoura spinetorum*) (Pl. 14a). UP dull blue; UN sepia brown, a thin irregular light line across both wings; tails present. Size 1–1¼ in. (25–31 mm). Flight period irregular; May–September in northern, April–August in southern California; two broods. Found in association with coniferous forest throughout the state; widely distributed but local in the Sierra Nevada and northern California; in the San Francisco Bay area known only from Mt. Diablo; in southern California known from the Transverse Ranges and eastern Mojave Desert ranges, where it flies in pinyon pine forest; also in San Diego County. Attracted to moisture, flowers, and warm pavement. Adults hilltop. *Early stages*: *egg* echinoid; *larva* yellow green with raised waxy orange patches and whitish bars tinged with magenta; *pupa* brown with darker markings. *Larval food plants*: dwarf mistletoes (*Arceuthobium* spp.) (Viscaceae), parasites of conifers.

Johnson's Hairstreak   (*Mitoura johnsoni*) (Pl. 14b). UN marked much like Thicket Hairstreak; size larger; UP rusty brown. Size 1¼–1⅜ in. (31–35 mm). Flight period midsummer; one brood. Openings and clearings in coniferous forest of Sierra Nevada from Mariposa County north through Cascades to British Columbia. Rare and local. *Early stages*: *egg* and *pupa* undescribed; *larva* yellowish green with hastate (triangular) markings that increase its resemblance to its food plants. *Larval food plants*: Douglas's Dwarf Mistletoe (*Arceuthobium douglasii*), parasitic on Douglas Fir; Pine Dwarf Mistletoe (*A. campylopodium*), parasitic on pines (both Viscaceae).

Moss's Hairstreak   (*Incisalia mossii*). UP male dark brown, female buffy; UN has a broad, irregular dark band; HW scalloped; tails absent. Size 1.0 in. (25 mm). Flight period March–April (coastal), April–June (inland); one brood. Found on rocky hills and outcrops. Three subspecies in California: (1) San Bruno Elfin, Bay Region Hairstreak (*I. m. bayensis*) (Pl. 14l), UN dark, contrasty; known only from San Bruno Mountains, San Mateo County. (2) Wind's Hairstreak (*I. m. windi*), UN lighter, nearly unmarked and uniform in color; west slope of the Sierra Nevada, and northward; usually scarce. (3) Doudoroff's Hairstreak (*I. m. doudoroffi*) (Pl. 14k), UN brown, dark markings complete but dull; coastal Monterey County from Big Sur south along coastal cliffs. *Early stages*: *egg* pale bluish green, laid singly on food plant; *larva* yellow or red, with light lateral chevrons; *pupa* brown. *Larval food plant*: Pacific Stonecrop (*Sedum spathulifolium*) (Crassulaceae).

Fotis Hairstreak   (*Incisalia fotis*) (Pl. 14j). UP uniform gray brown; UN distinctive, HW dark brown basally, light brown near outer border, these areas separated by a jagged dark line accented with white; a series of whitish scallops inside wing margin. Size ⅞–1 in. (22–25 mm). Flight period March–May; one brood. Flies in eastern Mojave Desert ranges of Inyo and San Bernardino counties east to Arizona and Utah, often in company with Comstock's Hairstreak. *Early stages*: undescribed. *Larval food plant*: Cliff Rose (*Cowania mexicana* var. *stansburiana*) (Rosaceae).

Western Brown Elfin   (*Incisalia augusta iroides*) (Pl. 14h). Brown, both UP and UN; small, turned-down projection at

anal angle of HW; tails absent. Size 1 in. (25 mm). Flight period February–April at low elevations; May–June in higher
mountains; one brood. Widely distributed; found in chaparral
and at forest edges. Eastern Mojave Desert populations resemble Annette's Brown Elfin (*I. a. annetteae*) (Pl. 14i). *Early
stages*: *egg* green; *larva* olive green, with raised triangular
area bordered by white bands; *pupa* brown, speckled with
black and with rows of black blotches on abdomen. *Larval
food plants*: dodder (*Cuscuta* spp.) (Convolvulaceae); California wild lilac (*Ceanothus* spp.) (Rhamnaceae); Madrone (*Ar-
butus menziesii*) (Ericaceae); Soap Plant (*Chlorogalum pome-
ridianum*) (Liliaceae); and many others. The larva eats mostly
the buds and flowers.

Western Banded Elfin   (*Incisalia eryphon*) (Pl. 14m).
UP male dark brown, female reddish brown; UN light brown,
irregularly banded with dark brown; HW scalloped; tails lacking. Size 1⅛–1¼ in. (28–31 mm). Flight period May–July;
one brood. Openings in pine forest; also in plantings of pine,
as near the Presidio in San Francisco. In southern California
limited to the higher mountain ranges, such as the San Bernardino Mountains near Big Bear Lake, and the San Jacinto
Mountains at Tahquitz Valley, both at 7,000 ft. Males perch on
or near pine trees, darting out to investigate possible mates.
Early stages: *egg* chalky white, flattened; *larva* green with
longitudinal cream stripes, hairy; *pupa* dark brown. *Larval
food plants*: various pines, including Lodgepole Pine (*Pinus
murrayana*) and Monterey Pine (*P. radiata*) (Pinaceae).

Bramble Hairstreak   (*Callophrys dumetorum*). UP male
gray, female rufous brown; UN yellow green; lower edge
UNFW broadly gray; a narrow dotted white line, edged with
black, across both wings. A turned-down lobe at HW anal angle; tails absent. Size 1–1¼ in. (25–31 mm). Flight period
March–May (June); one brood. Flies in wastelands, rocky hills,
chaparral openings in Coast Ranges, west slope Sierra Nevada
and Cascades, and in Transverse Ranges of southern California,
south into Baja California. Two subspecies: (1) Bramble Hairstreak (*C. d. dumetorum*) (Pl. 14n), as described above; north-

ern part of range. (2) Southern Bramble Hairstreak (*C. d. perplexa*), UNFW mostly gray, UNHW spot line greatly reduced, often absent; southern part of range. *Early stages*: *egg* green with network of white walls; *larva* green with dorsal and lateral yellow lines; *pupa* brown with black blotches, hairy. *Larval food plants*: wild buckwheat (*Eriogonum* spp.) (Polygonaceae); Deer Weed (*Lotus scoparius*) (Fabaceae).

Green Hairstreak   (*Callophrys viridis*) (Pl. 14o). UP both sexes gray; UN blue green; lower edge UNFW narrowly gray. Size 1⅛–1¼ in. (28–31 mm). Flight season March–April; one brood. Coastal hills of northern California, Monterey County to Mendocino County. Scarce and local in much of its range. *Early stages*: much like those of Bramble Hairstreak. *Larval food plant*: Wide-leaved Buckwheat (*Eriogonum latifolium*) (Polygonaceae).

Comstock's Hairstreak   (*Callophrys comstocki*) (Pl. 14q). UP gray; UN green, HW completely and FW incompletely crossed by a narrow irregular white line that often has black edging. Size ¹³⁄₁₆–¹⁵⁄₁₆ in. (21–23 mm). Flight season March–September; spring, summer, and fall broods depending on rainfall. Eastern Mojave Desert ranges of Inyo and San Bernardino counties. Scarce in collections, though locally abundant following rains, in its isolated range. Males are territorial. Adults visit flowers of Squaw Bush (*Rhus trilobata* var. *anisophylla*). *Early stages*: undescribed. *Larval food plants*: wild buckwheat (*Eriogonum* spp.) (Polygonaceae).

Lembert's Hairstreak   (*Callophrys lemberti*) (Pl. 14p). UP male brownish gray, female more buffy. UN yellowish green, color less vivid than that of Bramble Hairstreak; the narrow white line across wings UN often broken into separate spots, and faced with brown rather than black. Size ⅞–1¼ in. (22–31 mm). Flies over rocky slopes and ridges at high elevations in the Sierra Nevada, north into Oregon. *Early stages*: unknown. *Larval food plant*: Frosty Buckwheat (*Eriogonum incanum*), perhaps others (Polygonaceae).

Great Purple Hairstreak   (*Atlides halesus estesi*) (Pl. 14u).
A large, showy hairstreak, uncommon but widely distributed.
UP iridescent blue (not purple, despite its name); UNHW has
red-and-green markings. Size 1¼–1½ in. (31–38 mm). Flight
period March–October; successive broods. Early spring adults
from overwintered pupae. Flies in oak woodland and along
stream bottoms where cottonwoods, sycamores, and ash trees
grow; at times pupae may be gathered in numbers from litter
beneath such trees on which mistletoe grows. *Early stages*: *larva*
green, covered with short orange hairs; *pupa* brown with black
wing covers; body short, plump, and hairy. *Larval food plant*:
Common Mistletoe (*Phoradendron flavescens*) (Viscaceae).

Avalon Hairstreak   (*Strymon avalona*) (Pl. 14r). Smaller
than the Common Hairstreak, lacking the heavy black line UN,
and with the orange spot at base of tails reduced. The Avalon
Hairstreak is restricted to Santa Catalina Island off southern
California. The only endemic full species of butterfly on any
of the Channel Islands, it is a rarity in collections. (The other
Island endemics are subspecies.) Named for the town of Avalon, it is equally common at the Isthmus, where it flies on the
campus of the University of Southern California's Marine Science Institute. Size ¹³⁄₁₆–¹⁵⁄₁₆ in. (21–23 mm). Flight period
February–October; several broods. *Early stages*: *egg* bluish
green, laid singly on food plant buds; *larva* green or pink, mi-
nutely hairy; *pupa* brown or pink with darker mottlings. *Larval
food plants*: Deer Weed (*Lotus scoparius*), Silver-leaved Trefoil (*L. argophyllum* var. *ornithopus*) (Fabaceae).

Common Hairstreak   (*Strymon melinus pudicus*) (Pls.
1c, 1d, 14s). Mouse gray above; red spots at base of tails. UN
lighter gray, a narrow line of vein-to-vein black spots faced
with white, across both wings. Red spots at base of tails. Size
1–1¼ in. (25–31 mm). Flight period March–October in
northern, February–November in southern California; several
broods. Scarce in early season; often common in the fall.
Found throughout California from lowlands through chaparral
to middle elevations. Occurs on Santa Cruz Island; replaced by

Avalon Hairstreak on Santa Catalina Island. Rubs hind wings together when perched. *Early stages*: *egg* green, flattened; *larva* green, has multicolored markings on each segment; hairy; *pupa* light brown with darker mottling. *Larval food plants*: buds and young seeds of many plants, including mallow (*Malva* spp.) and rose mallow (*Hibiscus* spp.) (Malvaceae); Garden Bean (*Phaseolus vulgaris*), lupine (*Lupinus* spp.), and false indigo (*Amorpha* spp.) (Fabaceae); Hops (*Humulus lupulus*) (Moraceae); Buckwheat (*Eriogonum* spp.), knotweed and smartweed (*Polygonum* spp.) (Polygonaceae); and others.

Columella Hairstreak (*Strymon columella istapa*) (Pl. 14t). UP similar in color to that of Leda Hairstreak; size a bit larger, ⅞–1⅛ in. (22–28 mm). May be distinguished by the single pair of tails. Flight period August–October; multiple brooded. Found sparingly in southern California in the Coachella and Imperial valleys of Riverside and Imperial counties, and in the Anza-Borrego Desert of San Diego County. Columella is more at home in southern Arizona and southward. *Early stages*: *egg* bright green; *larva* dark green, a maroon-shaded whitish dorsal shield and many brown hairs arising from white dots; *pupa* pinkish buff, brown speckled. *Larval food plant*: Alkali Mallow (*Sida hederacea*) (Malvaceae); may be raised on mallow (*Malva* spp.).

Subfamily Lycaeninae

FW radius four branched, the last two veins arising on one branch (stalked); eyes bare; palpi slanting, third segment rather long, smoothly scaled. HW tornus evenly rounded except in *Tharsalea*, where it is lobed and tailed. Spines of underside of tarsi numerous, not arranged in rows. Coloration coppery, blue, or gray in males, mottled in females.

Tailed Copper, Arota Copper (*Tharsalea arota*). UP male dull coppery; female spotted; UN light brown with many small black spots, a light band near outer edge; a red spot at anal angle of HW; tailed. Size 1⅛–1⅜ in. (28–34 mm). Found along streamsides, at forest edges, and in sagebrush lands.

Flight period May–June (coast), June–July (mountains); one brood. Subspecies: (1) Arota Copper (*T. a. arota*), west of Sierra Nevada–Cascade crest and in a broad arc of southern California mountains from the Tehachapi Mountains to the Santa Ana Mountains, excluding the Santa Monica Mountains, where (2) Cloudy Copper (*T. a. nubila*) (Pl. 15d), darker, with wider marginal band UN, occurs from Malibu to Mt. Wilson. (3) Virginia Copper (*T. a. virginiensis*), UN pale, eastern California, Nevada. *Early stages*: *egg* turban shaped, ribbed, white; *larva* green with a double white line above and a yellow lateral line; hairy; *pupa*: yellowish brown to brown. *Larval food plants*: various species of currant and gooseberry (*Ribes* spp.) (Saxifragaceae).

American Copper   (*Lycaena phlaeas hypophlaeas*) (Pl. 15e). UPFW bright copper, spotted and bordered with brownish black; UPHW dark with copper border. UNFW much like UPFW; UNHW dull brown with small dark spots and irregular orange border. Size ⅞–1⅛ in. (22–28 mm). The Californian subspecies of this wide-ranging copper flies on the highest slopes and passes of the central Sierra Nevada in midsummer. Known localities few: Mt. Maclure, Mt. Dana, Mono Pass, at about 12,000 ft. *Early stages* and *larval food plant* of the Californian subspecies unknown. Of eastern subspecies, dock (*Rumex* spp.) (Polygonaceae).

Lustrous Copper   (*Lycaena cuprea*) (Pl. 15g). A truly bright copper, UP fiery copper red except for a black border and some small black spots; UN light gray; FW has basal copper flush; HW has small black spots and red marginal line. Size 1⅛–1¼ in. (28–31 mm). Flight period late June–early August; one brood. Flies in meadows, forest openings, and alpine fell-fields of Sierra Nevada from Tulare County north into Oregon, at moderate to high elevations. Adults are attracted to the blossoms of groundsel (*Senecio* spp.) (Asteraceae), and to Pussy-paws (*Calyptridium umbellatum*) (Portulacaceae). *Early stages*: unknown. *Larval food plant*: Alpine Dock (Polygonaceae).

Great Copper   (*Gaeides xanthoides*) (formerly *Lycaena*). UP male coppery gray, female spotted with black and pale orange; a pale orange marginal band on HW; UN dull cream color with many small black spots and a partial red line at anal angle of HW. Size 1⅜–1½ in. (34–38 mm). Flight period May–June, to July at higher elevations; one brood. Found in fields and meadows in Coast Ranges, and on west slope of Sierra Nevada north to Siskiyou Mountains; in Transverse Ranges of southern California south to San Diego County. Two subspecies: (1) Great Copper (*G. x. xanthoides*), described above; northern part of range. (2) Mourning-garbed Copper (*G. x. luctuosa*) (Pl. 15b), duller; UNHW marginal spots bordered by irregular white submarginal markings; southern part of range. *Early stages: egg* pale green, pitted; *larva* variously colored, pale green, dark green, or orange barred with magenta; *pupa* pinkish buff with brown blotches. *Larval food plants*: dock, sorrel (*Rumex* spp.), including Fiddle Dock (*Rumex pulcher*), Curly Dock (*R. crispus*), Wild Rhubarb (*R. hymenosepalus*) (Polygonaceae).

Edith's Copper   (*Gaeides editha*) (formerly *Lycaena*) (Pl. 15c). Resembles Great Copper; smaller, duller gray; dark spots UNHW show "raindrop" effect—open, enclosing a pale spot. Size 1⅛–1¼ in. (28–31 mm). Flight period late June–August; one brood. Flies in mountain meadows, forest openings, and alpine fell-fields of Sierra Nevada–Cascades, north to Washington, east into Rocky Mountains; not found in southern California. *Early stages*: Unrecorded. *Larval food plants*: Santa Rosa Horkelia (*Horkelia tenuiloba*), Dusky Horkelia (*H. fusca*), no doubt others (Rosaceae). Has oviposited on Alpine Dock (*Rumex paucifolius*) (Polygonaceae).

Gorgon Copper   (*Gaeides gorgon*) (formerly *Lycaena*) (Pl. 15a). UP male bright purplish copper, female brown with small black spots, yellowish dashes, and a pale orange marginal band on HW; UN yellowish gray, with many small black spots; HW has brick red spots along outer edge. Size 1¼–1⅜ in. (31–34 mm). Flight period May–June, in higher mountains to mid-July; one brood. Flies over hillsides, rocky out-

crops, cut banks, and roadsides at low elevations west of the
Sierra Nevada–Cascade crest and on coastal side of Transverse
Ranges of southern California south to Orange County. Adults
visit wild buckwheat flowers. *Early stages*: *egg* creamy white,
ribbed; *larva* pale green, hairy; *pupa* pale green shaded with
yellow, covered with minute mushroom-shaped projections.
Larval food plants: wild buckwheat (*Eriogonum* spp.), including Naked Buckwheat (*E. nudum*) in northern California and
Long-stemmed Buckwheat (*E. elongatum*) in southern California (Polygonaceae).

Ruddy Copper　　(*Chalceria rubida*) (formerly *Lycaena*)
(Pls. 15f, 15h). UP male bright reddish copper, female duller,
spotted; UNFW yellowish gray with black spots; UNHW
nearly white, sometimes with small black dots. Size 1⅛–1½
in. (28–38 mm). Flight period July–August; one brood. Flies
in fields, meadows, sagebrush, from Inyo County north to
British Columbia, east to Nebraska and the Dakotas; not found
in southern California. A little-known subspecies, the Monache Copper (*C. r. monachensis*), described in 1977, is found
in Monache Meadows, Tulare County. *Early stages*: unrecorded. *Larval food plants*: docks (*Rumex* spp.) (Polygonaceae). Although not usually an alpine species, the Ruddy Copper flies at Gaylor Lakes, above Tioga Pass, el. 10,500 ft.,
with typical Artemisian associates, the sagebrush being Roth-
rock's Sagebrush, (*Artemisia rothrockii*), not Great Basin
Sagebrush (*A. tridentata*) (Asteraceae).

Varied Blue, Blue Copper　　(*Chalceria heteronea*) (formerly *Lycaena*). UP male bright blue with narrow black
border; female dull, spotted; UNFW light with small black
spots; UNHW silky white, only occasionally with dark dots.
Size 1¼ in. (31 mm). Flight period May–June (coast), June–
early August (mountains); one brood. Scarce in Coast Ranges;
common in sagebrush in high Sierra Nevada and in eastern and
northern California, south to Tejon Mountains (Frazier Mountain Park); north to British Columbia, east to Rocky Mountains. *Subspecies*: (1) Varied Blue (*C. h. heteronea*) (Pl. 15k),
most of range. (2) Bright Blue Copper (*C. h. clara*) (Pl. 15i),

male blue a bit lighter; female dull blue with black spots and edges; Tejon (Mt. Pinos and vicinity), Tehachapi, and Piute mountains, Ventura County and Kern County. *Early stages*: apparently undescribed. *Larval food plants*: wild buckwheat (*Eriogonum* spp.) (Polygonaceae).

Purplish Copper (*Epidemia helloides*) (formerly *Lycaena*) (Pl. 15j). UP male dull copper with brilliant purple reflections and many small dark dots, dark border, and reddish submarginal line on HW; female markings vaguely similar but more spotted, ground color sometimes yellowish or buffy. UN orange brown, with black spots and reddish marginal spots. Size 1–1⅜ in. (25–34 mm). Flight period April–October; multiple broods. Common in fields, yards, vacant lots, and marshy areas, at low elevations; less common in meadows at higher elevations. *Early stages*: *egg* echinoid, flattened, covered with nodules; *larva* bright green with a yellow lateral stripe, hairy; *pupa* green with brown markings and bristly tubercles. *Larval food plants*: many members of the Buckwheat family (Polygonaceae), including dock, sorrel (*Rumex* spp.), and knotweeds (*Polygonum* spp.). In dry yards and vacant lots, Wire Grass, Yard Knotweed (*P. aviculare*); in marshy areas, Common Knotweed (*P. lapathifolium*); many others.

Nivalis Copper (*Epidemia nivalis*) (formerly *Lycaena*) (Pl. 15l). UP male coppery brown with lilac gloss; female brown, with pale orange dashes and small black spots, much as other female coppers; UNFW yellowish, with heavy black spots; UNHW basally green gray, outwardly pink or light brown, often unspotted in the Sierra Nevada but usually dark spotted in the Warner Mountains of Modoc County; variable. Size 1⅛–1¼ in. (28–31 mm). Flies in meadows, forest openings, along streamsides, and in sagebrush flats and alpine fellfields, from the Sierra Nevada north to British Columbia, east to the Rocky Mountains. Flight period June–August; one brood. *Early stages*: *egg* flattened, pitted, pale bluish; *larva* pale green, a claret dorsal line, with brownish hairs and white tubercles; *pupa* straw yellow, with brown spots. *Larval food plants*: Douglas's Knotweed (*Polygonum douglasii*) specifically recorded; probably others (Polygonaceae).

Mariposa Copper   (*Epidemia mariposa*) (formerly *Lycaena*) (Pl. 15m). UP male coppery brown; female light orange brown with dark spots and borders, or sometimes much darker; UNFW yellow with dark spots and border; UNHW gray with thin crescentic dark spots; no pink or red. Size 1⅛–1¼ in. (28–31 mm). Flight period July–August; one brood. Flies in forest openings and at meadow edges in the Sierra Nevada and north to southern Alaska, east to the Rocky Mountains, at middle and high elevations; does not occur in southern California. Adults visit flowers, including Hoary Aster (*Machaeranthera canescens*) (Asteraceae). *Early stages*: undescribed. *Larval food plants*: knotweeds (*Polygonum* spp.), including Douglas's Knotweed (*P. douglasii*) (Polygonaceae).

Hermes Copper   (*Hermelycaena hermes*) (formerly *Tharsalea*) (Pl. 15n). UP brown; center of FW yellow with a few large brown spots; HW all brown except for a small yellow line at base of tail; UNFW yellow, a few large dark spots; UNHW all yellow except for a few tiny dark dots; HW tailed. Size 1–1⅛ in. (25–28 mm). Flight period mid-May to mid-July; one brood. This highly unusual tailed copper occupies a restricted range in San Diego County, extending north to Fallbrook, east to Descanso, and south to Santo Tomas, Baja California. This is mesa country, considered prime for homes, and much of it has been lost to developers. Males are territorial. Adults visit buckwheat flowers. *Early stages*: *egg* white, pitted; *larva* light green with yellow-bordered dark green dorsal stripe, head yellow, body with white projections; *pupa* light blue green with pale yellow cast on abdomen. *Larval food plant*: Redberry (*Rhamnus crocea*) (Rhamnaceae), a departure from the food plants of other coppers.

Subfamily Polyommatinae (formerly Plebejinae)

Structure much as in Lycaeninae, but palpi with third segment shorter, often hairy, and tarsal spines few, usually arranged in rows, one on each side. HW tornus evenly rounded, a short tail only in *Everes*. Usually smaller than other Lycaenidae. Coloration usually blue in males, darker bluish or brownish in females.

Pygmy Blue   (*Brephidium exile*) (Pl. 17h). UP buffy, the wing bases blue; UN pale brown with wavy white bands, a row of iridescent spots on outer edge HW. Our smallest butterfly. Size ½–¾ in. (13–19 mm). Flies over salt marshes, alkali flats, shadscale scrub, weed patches. Very widely distributed in lowlands and deserts. Found in Death Valley and on Santa Cruz and Santa Catalina islands. Flight period February– October; multiple broods. *Early stages*: *egg* green with white network; *larva* green with white dorsal and yellow lateral lines; covered with brown tubercles; *pupa* yellowish brown with brown dots; hairy. *Larval food plants*: saltbush of many species (*Atriplex* spp.), including Shadscale (*A. canescens*), Quail Brush (*A. lentiformis*), Fat Hen (*A. hastata*), Australian Saltbush (*A. semibaccata*), Pickleweed (*Salicornia ambigua*), Lamb's Quarters (*Chenopodium album*) (Chenopodiaceae); Petunia (*Petunia parviflora*) (Solanaceae) also reported.

Marine Blue   (*Leptotes marina*) (Pl. 16a). UP male lavender blue; two dark spots at back edge of HW; female more spotted; UN pale brown with many narrow pale bands; two iridescent spots at anal angle of HW. Size 1–1⅛ in. (25– 28 mm). Scarce in central California; occasional at low elevations east of the Sierra Nevada; much more common in southern California, in deserts, lowland areas, and Channel Islands (Santa Catalina, Santa Cruz). Flight period March–October; multiple broods. *Early stages*: *egg* turban shaped, green with white network; *larva* green to brown with darker brown bands and stripes; *pupa* brown with gray wing cases and spots of darker brown. *Larval food plants*: various legumes, including Alfalfa (*Medicago sativa*), pea (*Lathyrus* spp.), rattleweed (*Astragalus* spp.), Deer Weed (*Lotus scoparius*), and others; has been reared on *Wisteria* (all Fabaceae); leadwort (*Plumbago*) (Plumbaginaceae).

Edwards's Blue (*Hemiargus ceraunus gyas*) (Pl. 16b). Resembles Reakirt's Blue, but lacks the row of black eyespots UNFW. Edwards's Blue occurs throughout the desert parts of southern California. Though said to be rarely found in the mountains, it has been taken on Mt. Palomar, on the South Fork of the Santa Ana River in the San Bernardino Mountains, and at Idyllwild in the San Jacinto Mountains. Size ¾–⅞ in. (19–22 mm). Flight period March–October; several broods. *Early stages*: *egg* light green, the top flattened; *larva* green to yellow, darker green bands on each segment; hairy; *pupa* green with dark dorsal line. *Larval food plants*: Honey Mesquite (*Prosopis glandulosa* var. *torreyana*), Screw-bean Mesquite (*P. pubescens*), Alfalfa (*Medicago sativa*) (Fabaceae); others.

Reakirt's Blue (*Hemiargus isola alce*) (Pl. 16c). HW shorter than that of our other blues; UP male light blue, female much darker; UN light gray; UNFW has a cross-band of white-ringed black spots; HW has two-three marginal iridescent spots. Size ⅞–1 in. (22–25 mm). Flight period April–October; two broods. Flies in brushy and open lands east of the Sierra Nevada crest and in desert ranges of southern California. More usual at lower elevations, but has been taken as high as Tioga Pass, el. 9,941 ft. *Early stages*: unrecorded. *Larval food plants*: buds and tender pods of Honey Mesquite (*Prosopis glandulosa* var. *torreyana*); no doubt other legumes (Fabaceae).

Northern Blue (*Lycaeides idas*). UP male bright blue; female dark with a wavy orange submarginal band across both wings; UN has a complete submarginal band of faintly iridescent spots across both wings; a small dark spot at outer end of each wing vein. Size 1–1¼ in. (25–31 mm). Flight period June–July; one brood. Many subspecies; three in California: (1) Lotis Blue (*L. i. lotis*), colors subdued; coastal peat bogs of Sonoma–Mendocino counties; much of former habitat greatly altered by man's activities; local and endangered. (2) Anna Blue (*L. i. anna*) (Pl. 16f), colors very bright; moist meadows of Sierra Nevada, southern Oregon, western Nevada. (3) Rice's

Blue (*L. i. ricei*), iridescent spots of UN greatly reduced; UNHW nearly white; Siskiyou Mountains north into Oregon. *Early stages*: unrecorded for California subspecies. *Larval food plants*: in Sierra Nevada, lupine (*Lupinus* spp.); in Mendocino County, apparently a species of bird's-foot trefoil (*Lotus* sp.); elsewhere, stated to be pea (*Lathyrus* spp.) and vetch (*Vicia* spp.) (all Fabaceae).

Melissa Blue   (*Lycaeides melissa*). Very similar to the Northern Blue. UP blue of male warmer; orange band of female wider, iridescent spots of UN more vivid; black terminal line at wing edge distinct, the black vein-ends not isolated. Size 1–1¼ in. (25–31 mm). Flight period mid-May to mid-July; one brood; in southern California, occasionally a second brood in October. Flies in woodland openings, river bottoms, moist or marshy land; east of Sierra Nevada and southcoastal southern California, Tehachapi, Tejon, San Jacinto and Palomar Mountains. California has two or three subspecies: (1) Orange-margined Blue (*L. m. paradoxa*) (formerly *inyoensis*, and at one time mistakenly called Lotis Blue) (Pls. 16g, 16j), as described; east of Sierra Nevada. (2) Friday's Blue (*L. m. fridayi*), lighter blue, female orange markings reduced; TL Mammoth; high Sierra Nevada (Sonora Pass, etc.), not now considered valid. (3) An unnamed coastal subspecies found in Riverside County (Hemet), San Diego County (Warner Hot Springs), and south. *Early stages*: said to be similar to those of the San Emigdio Blue. *Larval food plants*: rattleweed (*Astragalus* spp.); lupine (*Lupinus* spp.); in Owens Valley, Wild Licorice (*Glycyrrhiza lepidota*) (all Fabaceae).

Greenish Blue, Saepiolus Blue   (*Plebejus saepiolus*). UP male cold blue; dusky borders wide; dark bar at end of FW cell; female similar, or much more often, brown to very dark brown; UN white, light gray, or (female) dull brown with many small black dots. Size ⅞–1¼ in. (22–31 mm). Flies in forest openings, at streamsides, in meadows, and alpine fell-fields, from cool places along the coast to high elevations in many of the California mountain ranges; the range covers much of the cooler parts of North America. Often very common. Flight period April–August (coastal), June–September (montane); two or more broods. Subspecies: (1) Greenish Blue (*P. s. saepiolus*) (Pl. 16e), northern part of range. (2) Hilda Blue (*P. s. hilda*) (Pl. 16d), UP female brown with orange submarginal band or lunules; higher elevations in the mountain ranges of southern California. *Early stages*: undescribed. *Larval food plants*: clovers (*Trifolium* spp.), including Forest Clover (*T. breweri*), and (for *P. s. hilda*) Carpet Clover (*T. monanthum*), Cow Clover (*T. wormskioldii*), and Long-stalked Clover (*T. longipes* var. *atrorubens*); in northern California, the second brood on bird's-foot trefoil (*Lotus* spp.) (all Fabaceae).

San Emigdio Blue   (*Plebulina emigdionis*) (formerly *Plebejus*) (Pl. 16i). UP lilac blue at base, brown near margins; HW with faint orange submarginal band; brown and orange areas more extensive in female; UN white, with many small black spots; UNHW with submarginal row of orange and iridescent spots. Size ⅞–1⅛ in. (22–28 mm). Once considered single brooded (April–May); now known to have summer (June–July) and fall (August–September) broods as well. Named for Mt. San Emigdio in the Tejon Range, Kern County, this lovely blue is better known from the Mojave River, near Victorville. It extends north into Inyo County and west into Ventura County, on the dry side of the Coast Ranges. The streams along which its food plant grows are mostly intermittent. *Early stages*: *egg* echinoid; *larva* green, gray, or brown, spotted with black; *pupa* green or light yellow. *Larval food plant*: Shadscale (*Atriplex canescens*) (Chenopodiaceae).

Icarioides Blue   (*Icaricia icarioides*). UP male light bright blue; dark borders narrow, wing veins dark at tips; obscure dark marginal HW spots present in some subspecies. Female sometimes blue, but usually dark; UN light gray; HW markings extremely variable; sometimes a double row of small dark spots, or dark spots centered or ringed with white; sometimes nearly unspotted. A large blue, size 1⅛–1⅜ in. (28–34 mm). Flight period April–August, depending on elevation; one brood. Coastal and montane; not found in hot central valleys or deserts. Ten subspecies in California, of which one is extinct: (1) Icarioides Blue (*I. i. icarioides*), UP female, discs of both FW and HW usually blue; UNHW dark spots usually small, white ringed; Siskiyou Mountains south in Sierra Nevada to Kern County. (2) Fulla Blue (*I. i. fulla*), very pale in both sexes; UNHW spots white, the dark centers often lacking; high Sierra Nevada; TL Sonora Pass, Mono County. (3) Mintha Blue (*I. i. mintha*), closely resembling the extinct Pheres Blue; larger, somewhat darker, female duller; TL Marin County; extent of range northward not clear. (4) Helios Blue (*I. i. helios*), male darker, cold blue; UNHW lightly marked; northern Mendocino County, north into Oregon; TL Fawn Lodge, Trinity County. (5) Evius Blue (*I. i. evius*) (Pl. 16l), male very pale, the veins silvery; female UP outer edge often has tan or orange scales; mountains of southern California. (6) Morro Blue (*I. i. moroensis*), dark border of FW male unusually wide; female brownish; UNHW spots white, sometimes reduced; coastal San Luis Obispo County; TL Morro Beach. (7) Mission Blue (*I. i. missionensis*), male darker blue than most subspecies; female usually dull brown; UNHW usually has a double row of small dark spots; TL Twin Peaks, San Francisco; south to San Bruno Mountains; endangered. (8) Ardea Blue (*I. i. ardea*), male light blue, the dark borders indefinite; female light brownish; UNHW spots usually white, sometimes dark centered; eastern California through Nevada and Utah. (9) Pardalis Blue (*I. i. pardalis*) (Pls. 16h, 16k), large, very clearly marked; female brown, a few small dark UPHW marginal spots; UNHW usually with two complete bands of dark spots; coastal counties from Sonoma County south to Monterey County; best defined in Contra Costa and Santa Clara counties. (10) Pheres Blue (*I. i. pheres*), male light blue, female light brown, the wing bases blue; a light cell spot on each wing; UN white spots usually without dark centers, often indistinct; TL San Francisco; formerly found on sand dunes of western San Francisco and in the Presidio; now extinct. *Early stages* (of *I. i. evius*): *egg* delicate green with white projections; *larva* has three diagonal bands per segment; *pupa* green, the abdomen red with green blotches. *Larval food plants*: various perennial lupines (*Lupinus* spp.), including (for *I. i. evius*) Dense-flowered Lupine (*L. densiflorus*) and (for *I.*

i. missionensis) Silver Lupine (*L. albifrons*), Summer Lupine (*L. formosus*), and Lindley's Lupine (*L. versicolor*) (Fabaceae).

Shasta Blue   (*Icaricia shasta*) (formerly *Plebejus*) (Pl. 16m). UP male lilac blue with wide dark borders and dark cell bars, both FW and HW; female darker; UN gray with dark gray spots; aurora (colored marginal band of HW) yellow or buffy, enclosing four to six iridescent green spots. Size ⅞– 1 in. (22–25 mm). Flight period June–August; one brood. Flies over meadows, in forest openings, and in alpine fell-fields at high elevations in the Sierra Nevada north to Washington, east to Wyoming. Formerly recognized as a subspecies, *I. s. comstocki*, the Yosemite population is no longer so considered. *Early stages*: undescribed. *Larval food plants*: Alpine Lupine (*Lupinus lyalli*), clover (*Trifolium* spp.) (Fabaceae).

Acmon Blue   (*Icaricia acmon*) (formerly *Plebejus*). UP male pale lilac blue; black border very narrow; HW marginal spots black, separate, aurora pink; female dark brown, sometime shot with blue, aurora orange; UN dull white with many small black spots; HW aurora orange, marginal spots iridescent green. Size ¾–1 in. (19–25 mm). Flight period February–October at lower elevations, where blue-scaled females (form "*cottlei*") are first to emerge; multiple broods. Distribution general; common all season at low elevations; less common at high elevations and in desert regions. Found on Santa Catalina and Santa Cruz islands. Subspecies: (1) Acmon Blue (*I. a. acmon*) (Pls. 16n, 1a, 1b), described above. (2) Texas Blue (*I. a.* nr. *texana*), Providence Mountains, San Bernardino County, and eastward. *Early stages*: *egg* echinoid, pale green; *larva* hairy, yellow with green dorsal stripe; *pupa* brown, its abdomen green. *Larval food plants*: bird's-foot trefoil (*Lotus* spp.), including Deerweed (*L. scoparius*), Spanish Clover (*L. purshianus*), rattleweed (*Astragalus* spp.), clover (*Trifolium* spp.), and other legumes (Fabaceae); also wild buckwheat (*Eriogonum* spp.) (Polygonaceae).

Lupine Blue   (*Icaricia lupini*) (formerly *Plebejus*). Much like Acmon Blue; slightly larger, wings slightly longer, dark border wider UPFW, veins slightly darkened, aurora orange red; UN like that of Acmon Blue. Size ⅞–1¼ in. (22–31 mm). Flight period late spring–summer (April–July); one brood. In brushland, sagebrush, mixed woodland, alpine meadows. Widely distributed at low and intermediate elevations; also found higher. Subspecies: (1) Lupine Blue (*I. l. lupini*), described above; northern part of range. (2) Clemence's Blue (*I. l. monticola*) (Pl. 16o), large, males clear blue, females more blue than other *Icaricia*; may have whitish markings just inside dark FW border; coastal and inland ranges of southern California. (3) Skinner's Blue, Green Blue (*I. i. chlorina*), UP male greenish blue, female often brown; Tehachapi Mountains, Tejon Mountains west of Lebec, Kern, Ventura, and northern Los Angeles counties; uncommon. *Early stages*: undescribed. *Larval food plants*: perennial buckwheats (*Eriogonum* spp.), including California Buckwheat (*E. fasciculatum*) and Sulfur Flower (*E. umbellatum*) (Polygonaceae). In spite of its name, *I. lupini* does not feed on Lupine.

Veined Blue   (*Icaricia neurona*) (formerly *Plebejus*) (Pl. 17q). UP brown in both sexes; veins of FW of male, of both wings of female, orange; veins often broadening into a wide submarginal band, especially in female. Size ¾–1 in. (19–25 mm). Flight period May–August; two broods. This rare "blue" occurs in scattered colonies from the Mt. Pinos area north through the Tehachapi Mountains and southern Sierra Nevada, east along the San Gabriel and San Bernardino mountains. One such colony is found at Doble, near Baldwin Lake, San Bernardino Mountains, at 8,000 ft. Found on mountain tops or openings in coniferous forest, often nectaring on Pussy-paws (*Calyptridium umbellatum*) (Portulacaceae). *Early stages*: *egg* echinoid, green, with a network of fine white lines; *larva* green, banded with darker green, covered with fine white hairs; *pupa* green with gray wing covers. *Larval food plant*: Wright's Buckwheat (*Eriogonum wrightii* var. *subscaphosum* and var. *trachygonum*) (Polygonaceae).

Gray Blue   (*Agriades franklinii podarce*) (formerly *A. aquilo podarce*) (Pl. 17a). UP male a distinctive bluish gray; borders, FW cell bar, and marginal spots dusky; female markings similar but ground color more brownish; UN dark spots widely ringed or surrounded by dull white. Size ⅞–1⅛ in. (22–28 mm). Flight period June–August; one brood. Flies in wet subalpine meadows and alpine fell-fields of Sierra Nevada and northward; does not occur in southern California. *Early stages*: undescribed. *Larval food plant*: reported to be shooting star (*Dodecatheon* spp.) (Primulaceae) in California.

Eastern Tailed Blue (*Everes comyntas*) (not shown). Resembles Western Tailed Blue but smaller; male UPHW marginal dark spot in cell CU_2 usually capped with pink. Positive identification only by dissection of male genitalia. Size ⅞–1⅛ in. (22–28 mm). Moist places and marshy land, mostly east of the Sierra Nevada, including Owens Valley, but also in much of cismontane California, including coastal and interior valleys. Scarce or overlooked in California. *Early stages*: similar to those of Western Tailed Blue. *Larval food plants*: various clovers (*Trifolium* spp.), often cultivated kinds; bird's-foot trefoil (*Lotus* spp.) (all Fabaceae).

Western Tailed Blue (*Everes amyntula*) (Pl. 16p). UP male pale blue, female much darker, UN nearly white, dark spots small and few; HW has a single short tail; marginal spots small, that in cell Cu_2 of male seldom pink capped. Size 1–1¼ in. (25–31 mm). Flight period March–June (coastal), June–August (Central Valley and Sierra Nevada); two broods at lower elevations. Forest edges, roadsides, marshy areas. In southern California, on Channel Islands (Anacapa, Santa Catalina, San Miguel); absent from deserts. Found in most of California; north to Alaska, east to Minnesota. *Early stages*: *egg* echinoid, green; *larva* yellow to green, often with pink or maroon markings, hairy; *pupa* buff or gray with brown blotches. *Larval food plants*: wild pea (*Lathyrus* spp.); bird's-foot trefoil (*Lotus* spp.); vetch (*Vicia* spp.); in Coast Ranges often Giant Vetch (*Vicia gigantea*); in southern California, rattleweed (*Astragalus* spp.) (all Fabaceae). Newly hatched larvae enter pods and feed on maturing seeds.

Square-spotted Blue (*Euphilotes battoides*). UP male violet blue with narrow dark borders and dark HW marginal

spots; female dark with orange HW aurora; UN spots prominent, square; aurora orange red. Size ¾–1 in. (19–25 mm).
Flight period March–August, depending on elevation; one
brood. Eight distinct subspecies occur in California: (1) Square-
spotted Blue (*E. b. battoides*) (Pl. 17b), UN gray, spots large,
square, often touching; high elevations in Sierra Nevada; July.
(2) Intermediate Blue (*E. b. intermedia*), UP dark borders
wide, UN black spots small; closely resembles Dotted Blue;
northern California; June–July. (3) San Bernardino Blue (*E.
b. bernardino*), very small; UNHW white, the spots separate,
distinct; Coast Ranges and Transverse Ranges north to San Be-
nito County; May–June. (4) Glaucous Blue (*E. b. glaucon*)
(Pl. 17d), pale blue, male with dark wing borders narrow, often
double; UNHW spots small; east of Sierra Nevada, Mono
County north into Oregon; June–July. (5) Martin's Blue (*E. b.
martini*), UP female bluish, UNHW aurora red, wide; eastern
Mojave Desert of Inyo and San Bernardino counties, east into
Arizona; late March–April. (6) El Segundo Blue, Allyn's Blue
(*E. b. allyni*), small; male with partial pink aurora UPHW; UN
markings very distinct; TL El Segundo, Los Angeles County;
late May–June; local, endangered. (7) Bauer's Blue (*E. b.
baueri*), TL west side Gilbert Pass, 6,200 ft., Inyo County;
little known. (8) Comstock's Blue (*E. b. comstocki*), UN black
spots very small and distinct; TL Tehachapi, Kern County; extent of range not known. *Early stages* (of *E. b. bernardino*):
egg echinoid, white; *larva* pale green with chocolate brown
pattern; hairy; *pupa* chestnut brown, hairless; pupation occurs
in dried flowers on which larvae have fed. *Larval food plants*:
various species of wild buckwheat (*Eriogonum* spp.) (Polygonaceae).

Dotted Blue   (*Euphilotes enoptes*). UN spots small, especially on FW, as compared with the Square-spotted Blue. In
the Sierra Nevada, the Dotted Blue is very distinct; the male is
dull blue with wide dusky borders; both fly there at the same
time. Elsewhere, the Dotted Blue may closely resemble the
Square-spotted Blue, but flies later in the year after the Square-
spotted Blue is past. Size ¾–1⅛ in. (19–28 mm). Flight period June–October, depending on locality; one brood. Six subspecies occur in California: (1) Dotted Blue (*E. e. enoptes*)

(Pl. 17c), Sierra Nevada–Cascades, moderate to high elevations; June–August. (2) Bay Region Blue (*E. e. bayensis*), bright blue with very narrow dark borders, UN light gray, spots small, aurora bright; eastern and northern edges of San Francisco Bay; local; June. (3) Tilden's Blue (*E. e. tildeni*), smaller, paler; inner Coast Ranges south of San Francisco Bay; July–early September; usually scarce. (4) Dammers's Blue (*E. e. dammersi*) (Pl. 17e), UP male bright blue; UP female dark brown; aurora restricted; a dark smudge UNFW; Colorado and Mojave deserts; TL Snow Creek, Riverside County; September–October. (5) Smith's Blue (*E. e. smithi*), UP, male dark blue, female dark brown; aurora distinct; colors and markings bright and clear; immediate coast of Monterey County; endangered. (6) Langston's Blue (*E. e. langstoni*), UP, male pale blue, aurora indistinct; female more or less blue at wing bases; TL 1.6 mi. north of Mono-Inyo county line, Hwy. 395; little known. *Early stages* (of *E. e. dammersi*): *egg* blue green, top depressed; *larva* white with pinkish brown middorsal stripe and lateral markings, hairy; *pupa* chestnut brown, hairless. *Larval food plants*: wild buckwheats (*Eriogonum* spp.) (Polygonaceae), of which only the flowers are eaten.

Mojave Blue   (*Euphilotes mojave*) (Pl. 17f). Male pale blue, dark border narrow, aurora dull; female quite blue, unusual in *Euphilotes*. Size ¾–1 in. (19–25 mm). Flight period April–May; one brood. This small and inconspicuous blue flies in the hills surrounding the Mojave Desert of southern California, and on the desert itself from northern Los Angeles County eastward through Riverside and San Bernardino counties, north of the Transverse Ranges. Typical localities are Little Rock and Joshua Tree National Monument. *Early stages*: *egg* white, top depressed; *larva* yellow, heavily patterned with rose, hairy; *pupa* brown, tinged with yellow. *Larval food plants*: wild buckwheats (*Eriogonum* spp.), including Puny Buckwheat, Yellow Turban (*E. pusillum*), and Kidneyleaf Buckwheat (*E. reniforme*) (Polygonaceae).

Elvira's Blue   (*Euphilotes pallescens elvirae*) (formerly *E. rita elvirae*) (Pl. 17g). UP male violet blue, dark borders narrow; female brown, HW aurora orange; UN spots, especially on FW, large, tending to diffuse; aurora wide, orange. Size ¾–1 in. (19–25 mm). Flight period July–October; one brood. Flies in foothills surrounding western Mojave Desert from Walker Pass, Kern County, to Pearblossom, Los Angeles County, at moderate elevations (3,000–4,000 ft.). *Early stages*: *egg* pale green with raised network, the top depressed; *larva* ivory white with greenish tinge, middorsal and lateral markings pinkish brown; *pupa* green, changing to orange yellow. *Larval food plants*: wild buckwheat (*Eriogonum* spp.), including Nuttall Buckwheat (*E. microthecum*) and Flat-topped Buckwheat (*E. plumatella*) (Polygonaceae).

Small Blue   (*Philotiella speciosa*). Very small; front wings long and narrow; UP, male dull blue, the dusky borders wide; female dark; UN dull white, FW black spots rather large; HW has a few minute black dots. Size ⅝–⅞ in. (16–22 mm). Flight period April–June; one brood. Sierra Nevada foothills (rare), south through Havilah, Red Rock Canyon, and Randsburg, Kern County, to San Diego and Imperial counties; isolated populations in southern Coastal Ranges. Subspecies: (1) Small Blue (*P. s. speciosa*) (Pl. 17i); most of range. (2) Boharts' Blue (*P. s. bohartorum*), vicinity of Briceburg, Mariposa County (TL), and at Hume Lake, Fresno County; very scarce. *Early stages*: *egg* echinoid, laid singly on food plant; *larva* green with rose-colored dorsal band, hairy; *pupa* chestnut brown. *Larval food plants*: Punctured Bract (*Oxytheca perfoliata*) and Trilobia (*O. trilobata*); also Kidneyleaf Buckwheat (*Eriogonum reniforme*) (all Polygonaceae); for *P. s. bohartorum*, not known.

Sonoran Blue   (*Philotes sonorensis*) (Pl. 17j). Considered our most beautiful small butterfly. UP male silvery blue, two red spots on FW; female has red spots on both FW and HW; UN dull gray, two red spots on FW; fringes checkered. Size ⅞–1 in. (22–25 mm). Flight period February–April; one brood. Found in rocky canyons and outcrops where the food plants grow; local and colonial. Inner Coast Ranges to Santa Clara County; Sierra Nevada foothills north to Placer County; San Gabriel and adjacent canyons of Los Angeles County south to Baja California. Not known from the desert ranges. *Early stages*: *egg* pea green with a raised network, flattened above; *larva* green to rose; *pupa* brown. *Larval food plants*: rock lettuce, live-for-ever, or "wild hen-and-chickens" (*Dudleya* spp.), including *D. lanceolata* (Crassulaceae). The larva bores into the fleshy leaves, pupates in trash on the ground.

Silvery Blue   (*Glaucopsyche lygdamus*). UP male cold silvery blue; the black border narrow; female brown, sometimes blue at wing bases; UN gray, a single line of round black spots ringed with white across both wings. Size 1–1¼ in. (25–31 mm). Flight period March–early August, depending on elevation; one brood. Widely distributed. Four subspecies in California: (1) Behr's Blue (*G. l. incognitus*) (formerly considered to be *G. l. behrii*, in error) (Pl. 17l); northern California west of the Sierra Nevada–Cascade crest, often at low elevations. (2) Columbia Blue (*G. l. columbia*), female usually dull blue; black spots of UN smaller; high elevations, Sierra Nevada and northward. (3) Southern Blue (*G. l. australis*), female more blue at wing bases; black dots UNHW reduced but distinct; Coast Ranges of southern California from San Luis Obispo County to San Diego County, east to edge of deserts; Santa Cruz Island. (4) Palos Verdes Blue (*G. l. palosverdesensis*), (Pl. 17m), female brown above, blue of wing bases restricted, black dots UNHW haloed by white rings; windward slopes of Palos Verdes Peninsula, Los Angeles County; endangered. Populations found in eastern Mojave Desert ranges have not yet been named. *Early stages*: *egg* echinoid, flattened, with a raised white network; *larva* bright green with a magenta dorsal band, hairy; *pupa* brown with black dots. *Larval food plants*: lupine (*Lupinus* spp.), including White-whorl Lupine (*L. densiflorus*); wild pea (*Lathyrus* spp.), vetch (*Vicia* spp.), and other legumes (Fabaceae).

Arrowhead Blue   (*Glaucopsyche piasus*). UP male blue with wide dusky borders; fringes checkered; female duller; UN light gray, with a few round black spots; HW has white submarginal arrowhead markings. Size 1⅛–1¼ in. (28–31 mm). Flight period April–July; one brood. Flies in cool forested areas and woodland roadsides; uncommon. Coast Ranges north of San Francisco Bay, Sierra Nevada, Cascades, Siskiyous; southern California, higher elevations of Tejon and San Bernardino mountains, sparingly in mountains of San Diego County; north to British Columbia; east to Rocky Mountains. Two subspecies in California: (1) Arrowhead Blue (*G. p. piasus*, as described above; northern part of range. (2) Coastal Arrowhead Blue (*G. p. catalina*) (Pl. 17k), smaller, UN darker; now very scarce, restricted to valleys bordering the Los Angeles Basin. *Early stages*: *larva* greenish white with broken dark red bands, finely speckled with white and red. *Larval food plants*: lupine (*Lupinus* spp.), including Silver Lupine (*L. albifrons*), Interior Bush Lupine (*L. excubitus*), and Stinging Lupine (*L. hirsutissimus*); rattleweed (*Astragalus* spp.) also recorded for western Sierra Nevada (all Fabaceae).

Xerces Blue   (*Glaucopsyche xerces*) (Pl. 17n). UP male lilac blue; female brown with a slight bluish gloss; UN light gray with round white spots (typical *xerces*), or black spots with white rings around them (form "*polyphemus*"). Size 1–1¼ in. (25–31 mm). Flight period March–April; one brood. This beautiful species was once quite common on the sand dunes of the Sunset District of San Francisco. Last seen in 1941, it is now extinct, a victim of urban expansion. The Xerces Society, formed to foster butterfly conservation, is a living memorial to this species. *Larval food plant*: said to have been a species of bird's-foot trefoil (*Lotus* sp.) (Fabaceae).

Spring Azure, Echo Blue   (*Celastrina ladon echo*) (formerly *C. argiolus echo*) (Pl. 17o). UP male azure blue; female duller, with dusky borders and indistinct HW marginal spots. UN dull white with tiny dark dots; fringes checkered. Size 1–1⅛ in. (25–28 mm). Flight period February–July, depending on elevation; two, perhaps three, broods. Found nearly everywhere except in cultivated farmlands. A second subspecies, the Cinereous Blue (*C. l. cinerea*) (Pl. 17p), smaller and darker, is found in the eastern Mojave Desert ranges. *Early stages*: *egg* turban shaped, pea green; *larva* green to magenta; *pupa* brown. *Early stages* of Californian subspecies unrecorded. *Larval food plants*: buds, flowers, and green seed pods of many trees and shrubs, including Creek Dogwood (*Cornus occidentalis*) and Brown Dogwood (*C. glabrata*) (Cornaceae); California Buckeye (*Aesculus californica*) (Hippocastanaceae); California lilac (*Ceanothus* spp.) (Rhamnaceae); and oak (*Quercus* spp.) (Fagaceae); for the Cinereous Blue, Rock Spirea (*Petrophytum caespitosum*) (Rosaceae).

Giant Skippers (Family Megathymidae)

Large, 1¾–2⅝ in. (43–66 mm), heavy-bodied skippers of powerful flight, inhabiting desert regions of the southwestern U.S. and northern Mexico. *Larvae* burrow in pithy stems and roots of plants of the family Agavaceae (yuccas and century plants), where pupation takes place. *Pupae* move readily within larval excavations. Adults have narrow head, clubbed antennae, and sturdy wings; flight is swift and erratic. Of California's two genera, *Megathymus* occurs on *Yucca* and adults fly in the spring. *Agathymus* occurs on *Agave* and adults fly in the fall. One brood a year.

Stephens's Giant Skipper   (Stephens's Agave Borer) (*Agathymus stephensi*) (Pl. 18c). UP light brown, with pale overscaling basally; a submarginal row of pale yellow spots, larger in female than in male; fringes checkered. Size 1⅞–2¼ in. (47–57 mm). Flight period September–October; one brood. Flies along western fringe of the Colorado Desert from Riverside County (Palms-to-Pines Hwy.) through San Diego County (Sentenac Canyon), to Baja California; abundant in Anza-Borrego Park. *Early stages*: *egg* nearly spherical, laid at base of leaf of century plant; *larva* burrows into leaf and pupates there; *pupa* hatches and adult emerges through an exit prepared by the larva before pupation. Adults perch on nearby shrubs; may be seen most readily at night, when their eyes reflect the beam of the collector's lantern. *Larval food plant*: Desert Agave (*Agave deserti*) (Agavaceae).

Allie's Giant Skipper   (Allie's Agave Borer) (*Agathymus alliae*) (Pls. 18b, 24a). UP dark brown, with orange basal overscaling; submarginal row of yellow spots clearly defined on FW, these larger in female than in male; HW yellow spots blend with orange background. Size 2–2⅝ in. (50–67 mm). Flight period September–October; one brood. Flies in eastern Mojave Desert (Kingston, Clark, Ivanpah mountains) at 3,000–5,000 ft., and in adjacent southern Nevada and northern Arizona. *Early stages*: partially known; *egg* laid on leaves of medium-sized plants; *larva* burrows in leaf; *adult* emerges through a trapdoor on underside of leaf. *Larval food plant*: Pygmy Agave (*Agave utahensis* var. *nevadensis*) (Agavaceae).

Bauer's Giant Skipper   (Bauer's Agave Borer) (*Agathymus baueri*) (Pls. 18d, 24b). UP dark brown, male with basal orange overscaling obscuring HW submarginal row of yellow spots; female with orange predominating over brown; fringes checkered. Size 2–2½ in. (50–64 mm). Flight period September–October; one brood. Flies in Providence and Granite mountains of San Bernardino County over rocky slopes at 4,000–6,000 ft. *Early stages*: undescribed. *Larval food plant*: Desert Agave (*Agave deserti*) (Agavaceae).

Common Giant Skipper   (Colorado Yucca Borer) (*Megathymus coloradensis*). UP brownish black, lighter basally; a band of creamy yellow submarginal spots on FW; female smaller submarginal spots on HW also; FW fringes checkered. HW has yellow margin. Size 2–2½ in. (50–64 mm). Flight period March–April; one brood. Two quite similar subspecies occur in California: (1) Martin's Giant Skipper (Martin's Yucca Borer) (*M. c. martini*), western fringe of Mojave and Colorado deserts from Los Angeles County (Little Rock) through Riverside County to San Diego County in the *Yucca* belt. (2) Maud's Giant Skipper (Maud's Yucca Borer) (*M. c. maudae*) (Pl. 18a), eastern Mojave Desert ranges (Kingston, Clark, New York, Providence mountains). *Early stages* (of *M. c. martini*): *egg* hemispherical, deposited on new shoots of yucca plants; *larva* black collared, burrows to center of the small plant and builds a mound ("tent") of silk and droppings; *pupa* with caudal (tail) projection; capable of rapid movement when disturbed. *Larval* *food plants*: of *M. c. martini*, Joshua Tree (*Yucca brevifolia*) and Mojave Yucca (*Y. schidigera*); of *M. c. maudae*, both of above and also Flesh-fruited Yucca (*Y. semibaccata*), the preferred host (all Agavaceae).

Skippers (Family Hesperiidae)

Small to medium-sized butterflies, ¾–2¼ in. (19–52 mm), of robust build and rapid flight, reaching their greatest diversity in the New World tropics. Head wide, antennae far apart, basal segment with a tuft of hairs ("eyelash"). Wings short and stiff, those of subfamily Pyrginae usually held open upon alighting, those of subfamilies Hesperiinae and Heteropterinae usually held closed upright over back, or partly open. *Egg* sea-urchin shaped to hemispherical; *larva* with small prothorax, forming a constriction ("neck") between head and mesothorax; *pupa* usually enclosed in a loose cocoon, often attached by a thread. Adults usually dull colored; yellow, tawny, buffy, dun, black, or checkered.

Subfamily Hesperiinae

Mid-tibia spined, except in *Euphyes*. Antennal club usually short and stout, apiculus usually sharp, in some genera reduced or nearly lacking. Palpi not hairy, upturned against the face, third segment often short and slender. Males with a FW stigma in some genera. Coloration yellow, tawny, rufous, or brown, often with light spots or darker borders.

Wandering Skipper   (*Panoquina errans*) (Pls. 18e, 24c). UP gray brown; several small white spots in FW; HW unmarked; UN gray brown with yellow along veins; a complete median row of small light spots across both wings; FW pointed. Size 1–1⅛ in. (25–28 mm). Flight period July–September; one brood. Flies along coastline from Santa Barbara to San Diego in scattered colonies, frequently at river mouths. *Early stages*: *egg* spherical, white; *larva* green with darker green dorsal and yellow lateral stripes; *pupa* greenish brown. *Larval food plants*: Salt Grass (*Distichlis spicata*) and other grasses (Poaceae).

FIG. 30 Brazilian Skipper

FIG. 31 Nyctelius Skipper

Brazilian Skipper, Canna Leaf Roller   (*Calpodes ethlius*) (Figure 30). A large, 1¾–1⅞ in. (44–47 mm), brown skipper, with hyaline spots in FW and a row of four hyaline spots in center of HW. *Larval food plant*: canna (*Canna* spp.) (Cannaceae), a "domestic" species in the U.S., where canna is not native, but grown only as a cultivated flower. Three California specimens known, two from Los Angeles, reared in 1904, and one from Riverside, 1924.

Nyctelius Skipper   (*Nyctelius nyctelius*) (Figure 31). A medium-sized, 1⅜–1½ in. (34–38 mm), dull-colored skipper. UNHW has two dull dark bands, and distinctive small dark costal spot. One definite California record, El Cajon, San Diego County, October 1958 (Oakley Shields).

Eufala Skipper   (*Lerodea eufala*) (Pls. 18f, 24d). UP gray brown, several white specks in FW; UN mouse gray, the white specks repeated; FW pointed. Size 1–1⅛ in. (25–29 mm). Flight period July–October; most abundant in fall; two or three broods. Pastures, fields, marshes; in northern California most common in Sacramento–San Joaquin Delta and in the Central Valley; in southern California found in Colorado, Coachella, and Imperial valleys; also coastal portions of Los Angeles, Orange, and San Diego counties. *Early stages*: *egg* pale green; *larva* bright green with hair-tipped papillae, dorsal line dark green; *pupa* green with dark green dorsal stripe. *Larval food plants*: grasses, especially Bermuda Grass (*Cynodon dactylon*) (Poaceae).

Roadside Skipper   (*Amblyscirtes vialis*) (Pl. 24e). UP deep brown, two or three small light spots near FW tip; UNHW with broad, purplish gray shades; fringes checkered. Size ⅞–1 in. (22–25 mm). Flight period June; one brood. Widely distributed throughout the United States; uncommon in California, where it is found in the northern Coast Ranges and the west slope of the Sierra Nevada. *Early stages*: *egg* light green, hemispherical; *larva* pale green with dark green dots at bases of short hairs; head white with vertical reddish stripes; overwinters as a *pupa*. *Larval food plants*: grasses, including blue grass (*Poa* spp.) (Poaceae).

Dun Skipper   (*Euphyes ruricola*) (formerly *E. vestris*) (Pl. 18g). Both sexes deep brown above and below; stigma of male with rusty outline; female with two or three pale FW spots;

UNHW often with an indistinct pale band. Size 1–1¼ in. (25–31 mm). Flight period late May–early July; one brood. Flies in woodland meadows, bogs, grasslands, from Santa Cruz County north along the coast; also San Diego County (Flynn Springs); rare or absent in interior; widely distributed, north to Canada, east to New England. *Early stages*: *egg* pale green, encircling band and apex red; *larva* green with whitish hairs; *pupa* greenish white, the wing pads yellowish. *Larval food plants*: unknown for California; in eastern United States, sedges (*Cyperus* spp.) (Cyperaceae) and grasses (Poaceae).

Umber Skipper (*Paratrytone melane*) (Pl. 18h). UP, both sexes, rich umber brown; wing bases red brown, a row of small light spots on outer third of FW; UN purplish brown, the pale spots repeated; a pale band HW. Size 1⅛–1¼ in. (28–31 mm). Flight period March–June and July–October; two broods. Streamsides, clearings, trails, roadsides, at low elevations near the coast and on west slope of the Sierra Nevada. *Early stages*: *egg* hemispherical, netted; *larva* yellow green, a black middorsal line and black dots in rows, hairy; *pupa* light yellow with pink overcast, deepening to ochre, hairy except wing cases. *Larval food plants*: various grasses; Hair Grass (*Deschampsia caespitosa*) in the Santa Monica Mountains; Goldentop (*Lamarckia aurea*) in San Diego; may be reared on Bermuda Grass (*Cynodon dactylon*) (Poaceae).

Woodland Skipper (*Ochlodes sylvanoides*). UP male bright red brown, the tip of the stigma touching the dusky border; UN red brown to chocolate brown, with or without a pale UN band; female similar but lacking stigma. Size ⅞–1⅛ in. (22–28 mm). Flight period July–October; one brood over most of range; two broods on the Channel Islands, where the Santa Cruz Skipper occurs in April–June and again in August. Subspecies: (1) Woodland Skipper (*O. s. sylvanoides*) (Pl. 18l), late summer and fall throughout California at low to moderate elevations; absent from desert regions, but common east of Sierra Nevada. (2) Santa Cruz Skipper (*O. s. santacruza*), UNHW darker, chocolate brown with contrasting yellow band. Channel Islands; coastal Santa Cruz and San Mateo counties populations seem to belong here also. *Early stages*: *egg* creamy white; laid singly on food plant; *larva* yellowish buff with seven lengthwise dark stripes, head black, has long summer diapause; *pupa* creamy white to dull brown, pupal period short. *Larval food plants*: various grasses, especially rye grass (*Elymus* spp.) but also many others, including lawn grasses (Poaceae).

Meadow Skipper   (*Ochlodes pratincola*) (not shown). Very small and pale, the black stigma of male conspicuous; tawny, with slightly darker borders; UN pale reddish brown; UNHW band pale and scarcely discernible. Size ⅞–1 in. (22–25 mm). Flight period June; one brood. Known mostly from the Tehachapi Mountains, especially in the vicinity of Tehachapi, Kern County. Apparently quite scarce. *Early stages*: unknown. *Larval food plants*: presumed to be grasses (Poaceae). This insect was formerly miscalled *O. nemorum*; however, *nemorum* is a synonym of *agricola*, and so applies to the Farmer, not to the Meadow Skipper.

The Farmer   (*Ochlodes agricola*). Male UP dark, the wing bases red brown; small clear spots near tip of stigma on FW; female similar but lacks stigma; UNHW reddish with an indistinct pale band. Size ⅞–1 in. (22–25 mm). Flight period May–July; one brood. Flies only in spring and summer, not in fall. Common at streamsides, roadsides, forest edges, and meadows at low elevations throughout California. Subspecies: (1) The Farmer (*O. a. agricola*) (Pl. 18j), as described above. (2) The Verus Farmer (*O. a. verus*) (Pls. 18i, 18k), much lighter in color; Kern River and Tehachapi Mountains areas. *Early stages*: *egg* conical, ribbed; *larva* (first instar) white with small black dots arranged in lines; head black; *pupa* undescribed. *Larval food plants*: grasses (Poaceae).

Yuma Skipper   (*Ochlodes yuma*) (Pl. 18m). Bright yellowish orange UP and UN; dusky borders narrow in male, indefinite in female; male stigma narrow, black, prominent. Larger than other somewhat similar skippers; size 1⅛–1⅜ in. (28–34 mm). Flight period June and August–September; two

broods. Flies in moist lands, marshes, and oases, in the Sacramento–San Joaquin Delta, the Central Valley, east of the Sierra
Nevada south to Haiwee Reservoir, Inyo County, and east to
Utah and western Colorado. Very local in distribution. *Early
stages*: *egg* greenish white, hemispherical; *mature larva* pale
green, the head creamy with vertical brown stripes; lives in
rolled leaf of food plant; *pupa* brown, abdomen with double
bands of small dark dots. *Larval food plant*: Common Reed
(*Phragmites australis* [=*communis*]) (Poaceae).

Field Skipper, Sachem   (*Atalopedes campestris*) (Pl. 18o).
UP male orange brown, the borders dusky; known by the very
large FW stigma; female darker brown, FW with two small
clear (hyaline) spots. Size 1–1¼ in. (25–32 mm). Flight
period April–June and August–October; two broods, perhaps
more. Widespread in California, but seldom common; said to
be more often found in disturbed areas: fields, cut-over land,
meadows, vacant lots, yards, roadsides. *Early stages*: *egg*
hemispherical, greenish white; *larva* olive green with a dark
greenish brown middorsal line, head and first body segment
black; *pupa* blackish brown. *Larval food plants*: various
grasses, including Bermuda Grass (*Cynodon dactylon*), St.
Augustine Grass (*Stenotaphrum secundatum*), and lawn grasses,
such as Kentucky Blue Grass (*Poa pratensis*) (Poaceae).

Sandhill Skipper   (*Polites sabuleti*). UP both sexes
tawny, the borders dusky; male FW stigma black with a gray
shade below; UNHW has an irregular yellowish patch and yellowish veins. Size ⅞–1 in. (22–25 mm). Widely distributed.
California has three subspecies: (1) Sandhill Skipper (*P. s.
sabuleti*) (Pl. 18p), low elevations throughout California;
April–September, commonest in fall. (2) Tecumseh Skipper
(*P. s. tecumseh*) (Pl. 18n), smaller, more brightly marked; high
elevations in Sierra Nevada–Cascade Range; July–August.
(3) Chusca Skipper (*P. s. chusca*) (formerly *P. s. comstocki*)
(Pl. 18r), lighter, UP dark margins reduced, UN markings ob-
solescent; Mojave and Colorado deserts from Palmdale and
Blythe south to Scissors Crossing and Imperial Valley; July–
October. *Early stages* (of *P. s. chusca*): *egg* green, the surface

granular, netted; *larva* green, head black; *pupa* yellowish green with pink overcast; *Larval food plants*: Salt Grass (*Distichlis spicata*), Bermuda Grass (*Cynodon dactylon*); lawn grasses of several kinds; no doubt others (Poaceae).

Tawny-edged Skipper (*Polites themistocles*) (Pl. 18q). UP male dark brown with tawny space between black stigma and FW costa; female all brown except for two or three small FW spots; UN coloration much like that of UP. Size 1–1⅛ in. (25–28 mm). Flight period June–July; one brood. Uncommon in glades and forest openings in northeastern California; common in much of eastern United States and southern Canada. *Early stages*: *egg* pale green; *larva* purplish brown to chocolate, the head black with vertical white stripes; *pupa* dull whitish to pale brown, head dusky, wing pads greenish. *Larval food plants*: panic grass (*Panicum* spp.) (Poaceae).

Sonora Skipper (*Polites sonora*). UP fulvous, the borders grayish brown; male stigma wide, black; female with fulvous markings more diffuse; UNHW spot band yellowish white, the spots separate. Size 1–1¼ in. (25–31 mm). Subspecies: (1) Sonora Skipper (*P. s. sonora*) (Pl. 18s), mountains of central and northern California; rare south of Kern County except in San Bernardino Mountains near Barton Flat. Flight period July–September; one brood. (2) Dog-star Skipper (*P. s. siris*) (Pl. 18t), very different; UP chocolate brown with bright red brown markings; UNHW seal brown with ivory band. Flight period May–June; one brood. Cool, bleak grasslands of northern California, north to British Columbia. *Early stages*: *egg* light green, hemispherical; *larva* grayish green, the head black; *pupa* undescribed. *Larval food plants*: grasses; Idaho Fescue (*Festuca idahoensis*) suspected (Poaceae).

Uncas Skipper (*Hesperia uncas macswaini*) (Pl. 19g). UP tawny, the dusky borders wide; UNHW dusky brown, the pale buffy band very irregular. Somewhat resembles the Nevada Skipper. Size 1⅛–1⅜ in. (28–34 mm). Flight period June–July; one brood. Flies in exposed, grassy places at very high elevations in the White Mountains of Inyo and Mono counties and adjacent Nevada; scarce and local. *Early stages*: *egg* greenish white, hemispherical; only stage described. *Larval food plant*: Nevada Needlegrass (*Stipa nevadensis*) (Poaceae).

Comma Skipper   (*Hesperia comma*). Our most variable skipper geographically. California has no less than six well-defined subspecies: (1) Harpalus Skipper (*H. c. harpalus*) (Pl. 19a), 1¼–1⅜ in. (31–34 mm). UP male bright fulvous; female duller; UNHW band white, complete; sagebrush, mixed brushland; east of the Sierra Nevada–Cascades; June–October. (2) Yosemite Skipper (*H. c. yosemite*) (Pl. 19b), smaller, 1–1⅜ in. (25–34 mm); UP light fulvous; UNHW band of small separate pale spots; west slope of Sierra Nevada from Tulare County north to Tehama County; June–July. (3) Oregon Skipper (*H. c. oregonia*) (Pl. 19c), 1–1¼ in. (25–31 mm), the bright fulvous areas reduced by wide, dark borders and wing bases; UNHW golden brown, the band whitish to yellowish; northern California (Siskiyou County) and Oregon; June–July. (4) Dodge's Skipper (*H. c. dodgei*) (Pl. 19e), 1–1¼ in. (25–31 mm), UP deep fulvous; UNHW chocolate brown, the band white to creamy, strongly contrasting; humid coastal strip from Santa Cruz County north to Marin County; late July–October. (5) Tilden's Skipper (*H. c. tildeni*) (Pl. 19f), small, ⅞–1⅛ in. (22–34 mm), "washed-out" coloration; UNHW pale brown, the band pale, narrow, often indistinct; inner Coast Ranges from Lake County south to San Luis Obispo County; August–October. (6) Leussler's Skipper (*H. c. leussleri*) (Pl. 19d), 1⅛–1⅜ in. (28–34 mm), bright fulvous areas expanded, dark borders reduced, particularly in the male; UNHW band of larger spots in groups of two or three; Transverse and Peninsular ranges of southern California, south to Baja California; May–August. *Early stages*: these differ slightly among the various subspecies; in general: *egg* creamy white; *larva* yellow with paler markings; *pupa* dark brown, the wing cases lighter; has glaucous bloom in some subspecies. *Larval food plants*: various grasses; sub-specifically: Thurber Needlegrass (*Stipa thurberiana*) for *har*-

palus; Malpais Blue Grass (*Poa scabrella*) for *tildeni*; Red Fescue (*Festuca rubra*) for *dodgei*; rye grass (*Lolium* spp.) and brome grass (*Bromus* spp.) for *oregonia* (all Poaceae).

Nevada Skipper   (*Hesperia nevada*) (Pl. 19h). UP pale tawny, the gray brown borders grading gradually into the ground color; UN light brown, the UNHW spot band very irregular, the spots separate, the last spot displaced far inward. Size 1–1¼ in. (25–31 mm). Flight period June–August; one brood. Frequents open, rocky areas on east slope of Sierra Nevada and eastward; high sagebrush association, 7,000–11,000 ft. *Early stages*: only partly described; *egg* hemispherical, dull white; *larva* has black head with front brown and white. *Larval food plant*: Western Needlegrass (*Stipa occidentalis*) is preferred; also Squirrel-tail Grass (*Sitanion hystrix*) (Poaceae).

Miriam's Skipper   (*Hesperia miriamae*) (Pl. 19i). UP pale tawny with a slight sheen; borders gray brown; veins and HW anal area dark; UN gray, only the FW cell tawny; spots of band large, extending along the veins. Size 1⅛–1¼ in. (28–31 mm). Flight period late June–July; one brood. A scarce and very local skipper, found only on the highest peaks of the Sierra Nevada and in the White Mountains, Inyo-Mono counties and adjacent Nevada. *Early stages*: *egg* creamy white; only stage described. *Larval food plant*: Blue-stem Beard Grass (*Andropogon scoparius*) (Poaceae).

Lindsey's Skipper   (*Hesperia lindseyi*) (Pl. 19j). UP bright fulvous, dark borders narrow; UNHW light brown, the veins pale; spots of band large, irregular, extending along the veins; small dark points at end of each vein. Size 1⅛–1⅜ in. (28–34 mm). Flight period May–July; one brood. Widely distributed; Oregon south through cismontane California, often in Coast Ranges; southern California from Coast Ranges and Tehachapi Mountains to San Jacinto Mountains, Riverside County. *Early stages*: *egg* dull white, in some populations, laid on a lichen (*Usnea florida*) growing on trees and fence posts; it overwinters and hatches in the spring; *larva* brown with pale bars above and on sides, hairy; *pupa* pale brown, with a waxy bloom. *Larval food plants*: Blue Bunch Grass (*Festuca idahoensis*), and Oat Grass (*Danthonia californica*) (Poaceae).

Columbian Skipper   (*Hesperia columbia*) (Pl. 19k). UP bright tawny, the borders and wing bases dark; UNHW golden brown, the spot band short, curved, and silvery white, quite striking. Size 1⅛–1⅜ in. (28–34 mm). Flight period March–May, and September–October; two broods. Flies in chaparral, oak woodland, and adjacent grasslands of Coast Ranges from Oregon south to San Diego County. More common in inner Coast Ranges. *Early stages*: *egg* white, tinged with yellow; *larva* creamy yellow with lighter bars; *pupa* light brown with darker brown mottling, and like the *larva*, hairy. *Larval food plants*: June Grass (*Koeleria macrantha*), and Oat-grass (*Danthonia californica*) (Poaceae).

Pahaska Skipper   (*Hesperia pahaska martini*) (Pl. 19l). UP male light fulvous, dark borders wide; female bright fulvous; UNHW dark brown, spots of band large, distinct. Size large for this genus, 1⅛–1⅜ in. (28–34 mm). Flight period June–July and September–October; two broods. Eastern Mojave Desert ranges of southern California, east into southern Nevada and northern Arizona. Adults are partial to nectar of rabbit brush (*Chrysothamnus* spp.) (Asteraceae), a fall bloomer. *Early stages*: *egg* spherical, greenish white; *larva* cream to yellow; *pupa* undescribed, *Larval food plant* (in California): Fluff Grass (*Erioneuron pulchellum*) (Poaceae).

Yuba Skipper   (*Hesperia juba*) (Pl. 19o). UP bright fulvous, dark borders not grading into fulvous of wing, but separate and distinct; UNHW band broad, irregular, usually white. Size large, 1¼–1⅝ in. (31–41 mm). Flight period irregular; has been taken from May to October in various parts of its range; apparently no definite broods. Widely distributed in western United States; in California found in most hilly and mountainous areas west of the deserts. *Early stages*: *egg* spherical, white with pinkish tint; *larva* cream color, hairy;

pupa brown with light and dark patches, hairy. *Larval food plants*: various grasses, the species not stated (Poaceae).

Eunus Skipper   (*Pseudocopaeodes eunus*) (Pl. 19m). Bright orange brown to yellowish orange; male with narrow diagonal black FW stigma; vein ends usually dark; UN dull yellow, the veins narrowly dark. Size 1–1⅛ in. (25–28 mm). Flight season spring to fall; two or three broods. Flies in northern California in lower San Joaquin Valley and east of the Sierra Nevada north at least to the Mono Basin; in southern California, alkali flats from Owens Valley, Inyo County, south through the Mojave River at Victorville, San Bernardino County, to Scissors Crossing, San Diego County. *Early stages*: *egg* cream color, conical, ribbed; *larva* and *pupa*: undescribed. *Larval food plant*: Desert Salt Grass (*Distichlis spicata* var. *stricta*) (Poaceae).

Carus Skipper   (*Yvretta carus*) (Pl. 19n). UP yellowish brown basally, brown marginally; margins wide; scattered white spots of FW more prominent in female than in male; a single row of HW spots in both sexes. Size 1–1¼ in. (25–31 mm). Flight period August–September; one brood. Flies in eastern San Bernardino and Riverside counties along the Colorado River. Favors well-irrigated lands, as at Fertilla, near Blythe. More common in Arizona. *Early stages* and *larval food plant*: unknown.

Fiery Skipper   (*Hylephila phyleus*) (Pls. 1g, 1h, 20a). UP male fiery orange with irregular dark border; female much duller and darker; both sexes have orange line crossing the dark border near anal angle of HW; UNHW dull orange with small dusky spots, the anal fold dark. Size 1⅛–1¼ in. (28–31 mm). Flight period April–December; multiple broods. Common at low elevations in cultivated areas, such as lawns, yards, fence rows, edges of swamps and marshes, abandoned fields, and vacant lots; scarce in undisturbed backlands. Adults attracted to many kinds of flowers; very partial to *Lantana* spp. (Verbenaceae). *Early stages*: *egg* hemispherical, smooth, bluish green; *larva* yellowish brown with brownish stripes, the head black; *pupa* yellowish brown with a brown line down the back; hairy. *Larval food plants*: many kinds of grasses, including Bermuda Grass (*Cynodon dactylon*) and lawn grasses (Poaceae). A common lawn skipper.

Hewitson's Skipper   (*Copaeodes aurantiaca*) (Pl. 19p). UP male bright yellow, female orange yellow, FW faintly margined with brown; UN golden yellow. Size small, ¾–1 in. (19–25 mm). Flight period April–September; multiple broods. Flies in canyons and washes of the Mojave and Colorado deserts from the Panamint Mountains, Inyo County, to the Laguna Mountains, San Diego County. A typical locality is Chino Canyon near Palm Springs. *Early stages*: *egg* hemispherical, smooth, white; *larva* green with lengthwise purple lines; *pupa* straw colored, with a projecting beak enclosing the palpi. *Larval food plant*: Bermuda Grass (*Cynodon dactylon*) said to be preferred; no doubt others (Poaceae).

Julia's Skipper   (*Nastra julia*) (formerly *Lerodea*) (Pl. 24g). UP medium brown in both sexes, brown borders narrow, FW with several white flecks, HW without markings; UNHW dull rusty brown, very plain. Size 1–1¼ in. (25–31 mm). Flight period late August–October; one brood. Flies in eastern Colorado Desert of Imperial and Riverside counties, as at Blythe. A rarity in California. This species is separable from other *Nastra* by the rusty color UN. *Early stages*: undescribed. *Larval food plants*: unknown in nature; has been reared on St. Augustine Grass (*Stenotaphrum secundatum*) (Poaceae).

Neamathla Skipper   (*Nastra neamathla*) (Figure 32). Very similar to Julia's Skipper, but more fuscous, less rusty, especially UN. Size ⅞–1⅛ in. (22–29 mm). Found in Florida; and west in small numbers through Texas, New Mexico, and southern Arizona; reported to occur in southeastern California. Described before Julia's Skipper, to which early reports of it from California may very well apply. *Early stages* and *larval food plant*: unknown.

FIG. 32 Neamathla Skipper

FIG. 33 Pirus Skipperling

Subfamily Heteropterinae

Mid-tibia spined, as in Hesperiinae. Antennal club short, curved, apiculus lacking. Palpi long, hairy, thrust forward. Secondary sex characters, such as male FW costal fold or FW stigma, absent. Wings evenly rounded. Coloration brown or blackish with orange spots.

Pirus Skipperling   (*Piruna pirus*) (Figure 33). A small, 1⅛–1⅜ in. (28–34 mm), skipper; UP warm brown; several minute hyaline spots in FW; HW unmarked; UNHW rusty brown. One record, a fresh male taken September 2, 1934, by L. Hulbirt, a careful collector, near Carlsbad, San Diego County. A remarkable record, since this is a Rocky Mountain montane species. *Early stages* and *larval food plant*: unknown.

Arctic Skipper   (*Carterocephalus palaemon mandan*)
(Pl. 19q). Blackish with orange spots in both sexes; coloration
unique. Size 1–1¼ in. (25–31 mm). Flight period May–June;
one brood. Flies in undisturbed wet meadows and forest openings; known to visit flowers of *Iris* spp. (Iridaceae), often
perching head down. Definitely recorded from very few places
in northern California; among these, coastal Sonoma County,
where it is locally common, and Marin County, where it seems
to be rare. *Early stages*: *egg* greenish white, hemispherical;
larva glaucous green with dark green middorsal line, a pale
stripe on each side, and dark dots below; *pupa*; undescribed.
Larval food plant: in California, Purple Reed Grass (*Cala-
magrostis purpurascens*) (Poaceae).

Subfamily Pyrginae

Mid-tibia without spines. Antennal club curved, apiculus slender, in some genera recurved. Males of some species with a
FW costal fold, or a hind tibial tuft, or both. A few species
with long tails. Coloration brown, blackish, or checkered.

Common Sooty-wing   (*Pholisora catullus*) (Pl. 24h).
Sooty black above and below; FW has tiny white dots; HW
often has a band of white dots. Small, ⅞–1 in. (22–25 mm).
Flight period March–May in central California, April–September in southern California; one or two broods. Widely distributed in northern California, but uncommon and overlooked; in southern California found only where mountains
and desert meet, as at Little Rock, Los Angeles County, or
Jacumba, San Diego County. *Early stages*: *egg* cone shaped,
ribbed, yellow brown; *larva* yellowish green, tuberculate, the
tubercles hairy; *pupa* greenish yellow to brown; appears dusty.
Larval food plants: various pigweeds (*Chenopodium* spp.)
(Chenopodiaceae) and amaranths (*Amaranthus* spp.) (Amaran-
thaceae); ragweed (*Ambrosia* spp.) (Asteraceae) also reported.

Mojave Sooty-wing   (*Pholisora libya*) (Pls. 20b, 24i).
UP blackish, with a variable number of small, roundish white
spots. UNHW light gray, also white spotted. California's only

small dark skipper spotted with white both above and below. Size 1–1⅛ in. (25–28 mm). Flight period March–October; two or more broods. Inhabits the Mojave and Colorado deserts of southern California from Mono County and southern San Joaquin Valley to San Diego and Imperial counties. Two well-defined but unnamed populations occur, one in arid lands east of the Sierra Nevada in Inyo and Mono counties, the other in western San Joaquin Valley, to McKittrick and Coalinga, perhaps further north. *Early stages: egg* hemispherical, ribbed, orange to ivory white; *larva* bluish green with raised, hair-tipped points; head black; *pupa* brown, the wing cases black. *Larval food plants*: salt bushes (*Atriplex* spp.), especially Shadscale (*A. canescens*) (Chenopodiaceae).

Alpheus Sooty-wing   (*Pholisora alpheus oricus*) (Pl. 20f). UP black with considerable white overscaling; especially the female. Size 1–1⅛ in. (25–28 mm), larger than MacNeill's Sooty-wing, with which it was formerly confused. Flight period April–June; one brood. Mojave Desert of southern California, specifically Argus Mountains of Inyo County and Lucerne Valley of San Bernardino County, eastward to the Great Basin. A vigorous flier, as compared to the next species, *P. gracielae*. *Early stages*: unknown; the earlier description by Comstock applies to the following species. *Larval food plant*: Shadscale (*Atriplex canescens*) (Chenopodiaceae).

MacNeill's Sooty-wing   (*Pholisora gracielae*) (Pl. 20g). UP black, the white overscaling less conspicuous than in the Alpheus Sooty-wing; sexes quite similar. Size small, ¾–1 in. (19–25 mm). Flight period March–April and July–October; two broods. Colorado River margins in eastern San Bernardino, Riverside, and Imperial counties, specifically Parker Dam, Needles, and Blythe. A weak flier, MacNeill's Sooty-wing takes refuge inside the foliage of its food plant when disturbed. *Early stages* (as described by Comstock under the name *alpheus*): *egg* white, sculptured; *larva* green with white nodules, hairy; *pupa* straw colored, dusted with white. *Larval food plant*: Quail Brush, Lens-scale (*Atriplex lentiformis*) (Chenopodiaceae).

Erichson's Skipper   (*Heliopetes domicella*) (Pls. 20e, 24l). UP male resembles female of Large White Skipper, but white bands across wings wider; sexes similar. Size 1–1¼ in. (25–31 mm). Flight period September–October; one brood. Eastern San Bernardino County, in the vicinity of Parker Dam. Recently discovered in southern California, Erichson's Skipper is more common in Arizona and in Baja California. *Early stages*: unknown. *Larval food plant*: unknown, presumed to be members of the mallow family (Malvaceae).

Large White Skipper   (*Heliopetes ericetorum*) (Pls. 20c, 24j). Male white with dark-checkered edges; female gray-and-white checkered; UNHW white with rusty markings. Size large, 1¼–1⅜ in. (31–34 mm). Flight period March–October; two broods. Found in most of California in brushlands, clearings, openings in chaparral and forest; foothills to moderate elevations in Coast Ranges and Sierra-Cascades; scarce in parts of its range; common in much of southern California, including the desert ranges. Males patrol canyon bottoms. Adults attracted to flowers, especially California Yerba Santa (*Eriodictyon californicum*, and Thick-leaved Yerba Santa (*E. crassifolium*) (Hydrophyllaceae). *Early stages*: *egg* white, hemispherical, cross-ribbed; *larva* greenish yellow with green dorsal and lateral lines; hairy; *pupa* brown with two black spots on sides of thorax. *Larval food plants*: Desert Hollyhock (*Sphaeralcea ambigua*), Narrow-leaved Desert Hollyhock (*S. angustifolia*), Davidson's Bush Mallow (*Malacothamnus davidsoni*), Mesa Bush Mallow (*M. fasciculatus*), Desert Five-spot (*Eremalche rotundifolia*), Hollyhock (*Althea rosea*), weedy mallows (*Malva* spp.), and others (Malvaceae).

Laviana Skipper   (*Heliopetes laviana*) (Pls. 20d, 24k). UP white, FW apex and border dark; UNFW apex brown; UNHW light brown, the base darker, a diagonal light median line across the entire wing. Size 1⅜–1½ in. (34–38 mm). Flies in thorn forest openings and edges from Arizona to Texas, south to Argentina. In California known only from three specimens taken in the Joshua Tree National Monument, Riverside County, July 13, 1960. *Early stages*: unrecorded. *Larval food plants*:

Indian mallow (*Abutilon* spp.), Alkali Mallow (*Sida heder-acea*), globe mallow (*Sphaeralcea* spp.) (Malvaceae).

Little Checkered Skipper   (*Pyrgus scriptura*) (Pl. 24m).
UP white spots small, those of FW few; UNHW pale, the spots indistinct; male lacks costal fold on FW. Size small, ¾–1 in. (19–25 mm). Flight period March–October; multiple broods. Found in alkali flats, alkaline fields, usually at low elevations; in northern California most common in the Sacramento Delta region and in the interior valleys; in southern California found sparingly in the Mojave and Colorado deserts. *Early stages*: undescribed. *Larval food plant*: Alkali Mallow (*Sida hede-racea*) (Malvaceae).

Rural Skipper, Two-banded Skipper   (*Pyrgus ruralis*).
UP very dark gray, checkered with white; HW has two com-plete rows of linear white spots; male has FW costal fold. Size 1 in. (25 mm). Flight period March–June (Coast Ranges), June–July (Sierra Nevada); one brood. Flies in meadows, forest openings, forest edges. Two subspecies: (1) Rural Skip-per (*P. r. ruralis*) (Pl. 24n), described above; most of range. (2) Laguna Mountains Rural Skipper (*P. r. lagunae*), named by Scott in 1981; white markings extended, general appearance whitish; Laguna Mountains to Palomar Mountain, San Diego County in May–June; isolated population. *Early stages*: un-described. *Larval food plants*: Dusky Horkelia (*Horkelia fusca*) in Sierra Nevada; Bolander's Horkelia (*H. bolanderi* var. *clevelandii*) suggested for San Diego County (both Rosaceae). Reports of Checkerbloom (*Sidalcea malvaeflora*) (Malvaceae) are unverified.

Common Checkered Skipper   (*Pyrgus communis*) (Pl. 24o). UP male light gray, female dark gray, both with extensive checkered markings. Male has FW costal fold. Size 1–1⅛ in. (25–28 mm). Flight period extensive, all warm months of the year; multiple broods. Found in backyards, vacant lots, city parks, fields, cultivated lands, and along roadsides, in tempe-rate regions throughout the U.S. and Canada, south to Argen-tina. *Early stages*: *egg* hemispherical, pale green to cream;

larva whitish to brownish, a dark dorsal line and two whitish lines on each side; *pupa* light green anteriorly, brown posteriorly, with darker dots and streaks. *Larval food plants*: many members of the Mallow Family; most often, common mallows (*Malva* spp.) (Malvaceae).

Western Checkered Skipper   (*Pyrgus albescens*) (Pl. 24p). General appearance identical to that of the Common Checkered Skipper, from which it can be distinguished only by a constant difference in the male genitalia. (The Western Checkered Skipper has the tip of the valve stubby and rounded; in the Common Skipper this structure is slender and double). Size 1–1¼ in. (25–31 mm). Found in the lowest, hottest desert areas from San Diego County east to the Gulf Coast of Texas, south into Mexico. No differences in *early stages* or *larval food plants* have been noted. Previously considered a synonym or subspecies of the Common Checkered Skipper, the Western Checkered Skipper is currently considered a separate species.

Dreamy Dusky-wing   (*Erynnis icelus*) (Pl. 20h). FW short, wide; UP with no clear (hyaline) spots. FW has two indefinite dark bands, the space between them lighter. Male has a hair tuft on hind tibia. Size small for this genus, 1⅛–1¼ in. (28–31 mm). Flight period May–early June; one brood. Flies in woodland openings, clearings, trails, streamsides. Sierra Nevada–Cascade Range from Mariposa County north to Siskiyou and Modoc counties, north to British Columbia, east to New England. Found at intermediate elevations; uncommon in California. *Early stages*: *larva* green, the prothorax yellow, dorsal line dark, lateral line pale. *Larval food plants*: willow (*Salix* spp.), poplar, cottonwood (*Populus* spp.), Quaking Aspen (*P. tremuloides*) (Salicaceae).

Sleepy Dusky-wing   (*Erynnis brizo*). Blackish brown above and below; UP has two obscure incomplete black bands on each FW; no hyaline spots; often a submarginal row of obscure light specks on HW; male has no hair tuft on hind tibia. Size moderate, 1¼–1⅜ in. (31–34 mm). Flight period March–June, depending on locality; one brood. Flies over brushland, chaparral. California has two subspecies: (1) Wright's Dusky-wing (*E. b. lacustra*) (Pl. 20i), Coast Ranges north to Lake County, local or scarce; Transverse Ranges of southern California from northern Kern County south to San Diego County. (2) Burgess's Dusky-wing (*E. b. burgessi*) (Pl. 20j), eastern Mojave Desert ranges (Providence and New York mountains) in May; east to Colorado and west Texas. *Early stages*: undescribed. *Larval food plants*: oaks of several species over its entire range; believed to be Leather Oak (*Quercus durata*) in northern California (Fagaceae).

Persius Dusky-wing   (*Erynnis persius*) (Pl. 20k). Dusky black with a few small hyaline spots on FW; FW hairs erect, giving a rather hairy appearance. A very difficult species to recognize. Size 1⅛–1¼ in. (28–31 mm). Flight season May–July in California; one brood. Found in mountains, where it is generally distributed and not uncommon; local in the Sacramento Valley. *Early stages*: *larva* pale green, sprinkled with white dots, each bearing a short white hair; a thin dark dorsal line and variable yellowish lateral lines; head yellowish brown to black, with small light spots and stripes; *egg* and *pupa* apparently undescribed. *Larval food plants*: in Santa Cruz Mountains, associated with Black Cottonwood (*Populus trichocarpa*) (Salicaceae); in Arizona, False Lupine (*Thermopsis pinetorum*) (Fabaceae). It is possible that two or more species may be represented under the name *persius*, but present knowledge does not permit their separation.

Afranius Dusky-wing   (*Erynnis afranius*) (Pl. 20l). UP, male black with a few hyaline spots on FW; scaling very flat and smooth, without the erect hairs so noticeable in the Persius Dusky-wing; female similar, hyaline spots somewhat larger. Size 1⅛–1¼ in. (28–31 mm). Flight period March–August; two broods. Flies in Transverse Ranges of southern California at moderate elevations, from northern Los Angeles County and western San Bernardino County southward. *Early stages*: *egg* white to creamy yellow; ribbed; *larva* pale green, a darker dorsal and a yellow lateral line, hairy; *pupa* bright green with a black spiracle on thorax. *Larval food plant*: Spanish Clover (*Lotus purshianus*) (Fabaceae) in southern California.

Funereal Dusky-wing   (*Erynnis funeralis*) (Pl. 20m). Like
the Mournful Dusky-wing, black, the outer edge of HW white,
but FW long and narrow, a brown spot near FW tip. Male has
metatibial tuft. Size large, 1¼–1⅜ in. (31–34 mm). Flight
period February–October; three broods. Flies at low elevations in the Central Valley, inner Coast Ranges, Sacramento
Delta region, and Sierra Nevada foothills in northern California; widely distributed in southern California lowlands and
deserts. *Early stages*: *egg* hemispherical, ribbed, changing
from white through green and orange to brown; *larva* yellowish green, hairy; *pupa* green clouded with yellow. *Larval food
plants*: Deer Weed (*Lotus scoparius*), Alfalfa (*Medicago sativa*), Desert Ironwood (*Olneya tesota*), Colorado River Hemp
(*Sesbania exaltata*), and others (Fabaceae).

Pacuvius Dusky-wing   (*Erynnis pacuvius*). Nominate Pacuvius Dusky-wing, size 1⁵⁄₁₆–1½ in. (33–38 mm), found in
the Rocky Mountains, is much more contrastingly marked than
any of the three subspecies found in California, which are:
(1) Artful Dusky-wing (*E. p. callidus*) (Pl. 20n), brownish
black, lacking hyaline spots, FW slightly mottled; Coast Ranges
from Monterey County south to Baja California; Transverse
Ranges from Tehachapi Mountains, Kern County to Laguna
Mountains, San Diego County; May–July; in San Diego
County, April–October; two broods. (2) Grinnell's Dusky-wing
(*E. p. pernigra*) (Pl. 20o), quite black, the darkest of our dusky-
wings; Coast Ranges from Santa Cruz County to Sonoma
County, perhaps further; May–June; usually uncommon.
(3) Dyar's Dusky-wing (*E.p. lilius*) (Pl. 20p), brownish black;
hyaline spots well developed; FW somewhat mottled; Sierra
Nevada–Cascades north to Modoc County; June–July. *Early
stages*: undescribed. *Larval food plants*: various species of
California lilac, including (for Dyar's Dusky-wing) Snow Brush
(*C. cordulatus*) and (for Artful Dusky-wing) Hairy Ceanothus
(*C. oliganthus*) (Rhamnaceae).

Mournful Dusky-wing, Sad Dusky-wing   (*Erynnis tristis*)
(Pl. 20q). HW hind edge white as in Funereal Dusky-wing; a
bit smaller, FW shorter, FW hyaline spots well developed; no

brown spot in FW; male has no metatibial tuft. Size 1⅛–1¼ in. (28–31 mm). Flight period March–October; three broods at lower elevations. Flies in oak woodland, forest clearings, roadside oaks; sometimes parks and yards. In northern California common in San Francisco Bay region, inner valleys, and Sierra Nevada foothills; in southern California found in Coast Ranges from San Luis Obispo County south to San Diego County, and Transverse and Peninsular ranges from Tehachapi Mountains, Kern County, to Laguna Mountains, San Diego County. *Early stages*: *egg* spherical, ribbed, yellow to orange; *larva* grayish green with yellow lateral stripe, covered with mushroomlike white dots; *pupa* greenish gray, wing cases darker, hairy. *Larval food plants*: oaks of several species; Coast Live Oak (*Quercus agrifolia*), Blue Oak (*Q. douglasii*), Valley Oak (*Q. lobata*), also introduced Cork Oak (*Q. suber*) (Fagaceae).

Propertius Dusky-wing   (*Erynnis propertius*) (Pl. 20r). Dusky blackish brown; heavily scaled with grizzly gray; FW hyaline spots usually well developed, especially in the female. Our largest and commonest dusky-wing. Size 1⅜–1½ in. (34–38 mm). Flight period April–July in northern part, March–June in southern part of range; one brood. Hills and mountains throughout California in Coast and inland ranges; absent from desert ranges. *Early stages*: undescribed. *Larval food plants*: oaks of several species, including Coast Live Oak (*Quercus agrifolia*) and Oregon Oak (*Q. garryana*) (Fagaceae).

Powdered Skipper   (*Systasea zampa*) (Pl. 20x) (formerly known as *S. evansi*, a synonym of *zampa*; before 1941 considered identical with *S. pulverulenta*, a Texas species). UP olive gray, with markings best noted in the illustration; the FW has a white zig-zag line bordered with dark brown inwardly; wing margins irregular; FW has one indentation; HW has two. Size 1⅛–1⅜ in. (28–34 mm). Flight period February–April and September; two broods. Flies along western margins of Colorado Desert in Riverside and San Diego counties, as at Chino Canyon and Scissors Crossing, and into Baja California. More common in Arizona. *Early stages*: unknown. *Larval food plants*: none definitely known; Rock Hibiscus (*Hibiscus denudatus*) (Malvaceae) is suspected.

Ceos Sooty-wing (*Staphylus ceos*) (formerly *Pholisora*) (Pl. 24f). UP intense black; white scaling lacking; FW short; UPFW has tiny paired white spots at tip; top of head and palpi yellow. Small, 1–1⅛ in. (25–29 mm). Flight period in California, June so far as now known. Uncommon in the Imperial Valley of southern California. More common in Arizona, where it is on the wing from March to August. *Early stages*: unknown. *Larval food plant*: unknown.

Northern Cloudy-wing (*Thorybes pylades*) (Pl. 20s). Dark brown (not black); several very small hyaline spots in FW; UNHW dull brown with two irregular transverse dark bands. Largest of our three cloudy-wings, and the most widely distributed. Size 1⅜–1⅝ in. (34–41 mm). Flight period May–July; one brood. References in the literature to *T. mexicanus* and *T. diversus* from southern California apply to *T. pylades*. Distribution in California quite general, except for the very highest elevations, and the low deserts. *Early stages*: *egg* spherical, ribbed, white; *larva* orange buff with maroon blotches and maroon dorsal and lateral lines, covered with buff points each tipped by a hair; *pupa* dark brown, wing cases buff, hairy. *Larval food plants*: many legumes, including clover (*Trifolium* spp.), Alfalfa (*Medicago sativa*), rattleweed (*Astragalus* spp.), bird's-foot trefoil (*Lotus* spp.), and California False Indigo (*Amorpha californica*) (Fabaceae).

Diverse Cloudy-wing (*Thorybes diversus*) (Pl. 20u). UP dull gray brown; hyaline spots more numerous and placed in short diagonal rows. Smaller than Northern Cloudy-wing. Size 1¼–1⅜ in. (31–34 mm). Flight period June–July; one brood. Habitat restricted to forest openings and glens in main coniferous forest at moderate elevations, as at Mather, near Yosemite National Park. Known mostly from the west slope of Sierra Nevada–Cascades. *Early stages*: *egg* turquoise, changing to green, then to gray; *larva* dark olive green, with numerous small pale dots, a dark dorsal line and two pale lateral lines on each side. *Larval food plant*: Cow Clover (*Trifolium worm-skjoldii*) (Fabaceae).

Mexican Cloudy-wing   (*Thorybes mexicana*). UP medium brown; FW has fairly large hyaline spots, each with a dark boundary; UNHW light brown, flecked with darker brown; transverse bands indistinct. The smallest cloudy-wing, size 1⅛–1¼ in. (28–31 mm). Flight period June–August, depending on elevation; one brood. Formerly all California populations of this species were placed under *nevada*, but recent work indicates that three subspecies are found in California: (1) Nevada Cloudy-wing (*T. m. nevada*) (Pl. 20t), high elevations in the Sierra Nevada–Cascades, in California, western Nevada, and Oregon. (2) Emily's Cloudy-wing (*T. m. aemilea*), more boldly marked, brighter UN. Long included under *nevada*. Raised from synonymy by Shapiro (1981), who found it common in the Trinity Alps; its further range awaits discovery. (3) White Mountains Cloudy-wing (*T. m. blanca*), UN light striations whiter, wing fringes also lighter. Described (1981) from the White Mountains of eastern California and western Nevada, 7,500 ft. to above timberline. *Early stages*: undescribed. *Larval food plants*: mostly unknown; *T. m. nevada* was found to oviposit on Carpet Clover (*Trifolium monanthum*) (Fabaceae).

Long-tailed Skipper   (*Urbanus proteus*) (Pl. 20v). UP dark brown, glossed with iridescent green near wing bases; FW with row of four squarish transparent spots across the middle, and a few scattered spots near apex; HW with scalloped margin and a broad, outcurving tail. Size large, 1⅝–1⅞ in. (41–47 mm). Flight period June–November; emergence irregular; multiple broods. In California, flies along western margins of Colorado Desert, thence sporadically to the coast, from Long Beach south; southern U.S. south to Argentina. *Early stages*: *egg* light yellow; *larva* greenish with a dark green dorsal and yellow lateral stripes; head brown; *pupa* brown, dusted with white. *Larval food plants*: Honey Mesquite (*Prosopis glandulosa* var. *torreyana*), wisteria (*Wisteria* spp.), and various other legumes; in California, usually Garden

Beans (*Phaseolus vulgaris*) (all Fabaceae). The larva has at times been called the Bean-leaf Roller.

Simplicius Skipper   (*Urbanus simplicius*) (Pl. 20z). All brown above and below; fringes not checkered; tail long. The basal dark band UNHW extends straight to the inner costal spot and joins it. (In related species this band extends between the two costal spots.) The UNFW apex has a small dark smudge (related species have an irregular pale spot or a small dash). Size 1⅝–1⅞ in. (41–47 mm). Flight period irregular; multiple broods in tropics. *Early stages*: undescribed. *Larval food plant*: unknown in nature; has been reared once on Garden Bean (*Phaseolus vulgaris*) (Fabaceae) in San Diego County, so far the only California record.

Arizona Skipper   (*Polygonus leo arizonensis*) (Pl. 20w). UP brown, FW with three large transparent spots arranged in a triangle and three small spots in a line near the tip; HW has two indistinct dark bands; anal angle square cut, lobed but not tailed. UN markings somewhat similar; UNHW violet gray, the bands brown. Size large, 1⅝–1⅞ in. (41–47 mm). Flight period August–September; one brood. Found sparingly in Mojave and Colorado deserts and west lowlands of Los Angeles County (Glendora) and San Diego County (La Mesa), following desert winds. Adults partial to flowers of Lantana (*Lantana camara*) (Verbenaceae). *Early stages*: egg green, changing to reddish; *larva* yellowish green with inconspicuous fine white pile, lateral line yellow, below this a series of yellow blotches. *Larval food plants*: in Florida, Florida Fishpoison-tree (*Piscidia piscipula*) and others (Fabaceae); not known for California.

Silver-spotted Skipper   (*Epargyreus clarus*). UP dark brown; FW with diagonal row of large dull yellow spots; UNHW with large silver spot. Size large, 1¾–2¼ in. (44–54 mm). Flight period May–July; in southern California a second brood August–September. California has three subspecies: (1) Silver-spotted Skipper (*E. c. clarus*), central spot of yellow FW band large, overlapped by relatively large spot below it; UNHW silver spot large, its margins irregular; south-

ern California except San Diego County. (2) California Silver-
spotted Skipper (*E. c. californicus*) (Pl. 20y), much smaller
overall, lower spot of yellow FW band small, not overlapping
central spot; UNHW silver spot somewhat smaller, its upper
half very narrow; Coast Ranges from Bay region northward,
often associated with black locust; also Sierra Nevada foothills
where black locust persists in old mining areas. (3) Arizona
Silver-spotted Skipper (*E. c. huachuca*), largest subspecies,
lower spot of yellow FW band very small, not overlapping central spot, UNHW silver spot narrow at apex, wide below, margins usually smooth; San Diego County eastward through Arizona to New Mexico and southern Colorado. *Early stages*: *egg*
hemispherical, ribbed, green with a red spot on top; *larva* yellow green with dark cross-bands, the head brown with orange
spots; *pupa* brown, dusted with white. *Larval food plants*:
many legumes, including *Wisteria* spp., false indigo (*Amorpha* spp.), Broad-leaved Lotus (*Lotus crassifolius*); in northern
California the most common food plant is the cultivated Black
Locust (*Robinia pseudo-acacia*); in Arizona, New-Mexican
Locust (*R. neomexicana*) (all Fabaceae).

CHECKLIST OF
CALIFORNIA BUTTERFLIES

*(asterisk) indicates nominate subspecies does not occur in
  California.
†(dagger) indicates species or subspecies is extinct.

Family SATYRIDAE　　　　　Satyrs, Arctics, Ringlets

　1. *Coenonympha california*　California Ringlet
　　　Westwood
　　a. *california* Westwood
　　b. *eryngii* Hy. Edwards　　Siskiyou Ringlet

 *2. *Coenonympha ampelos*　　Ringless Ringlet
　　　W. H. Edwards
　　a. *elko* W. H. Edwards　　Ringless Ringlet

 *3. *Coenonympha ochracea*　Ochraceous Ringlet
　　　W. H. Edwards
　　a. *mono* Burdick　　　　Mono Ringlet

　4. *Neominois ridingsii*　　Riding's Satyr
　　　(W. H. Edwards)

 *5. *Cercyonis pegala*　　　Wood Nymph
　　　(Fabricius)
　　a. *boopis* (Behr)　　　　Ox-eyed Satyr
　　　form "*baroni*"　　　　Baron's Satyr
　　　(W. H. Edwards)
　　　form "*incana*"　　　　Hoary Satyr
　　　(W. H. Edwards)
　　b. ariane (Boisduval)　　Ariane Satyr
　　　form "*stephensi*"　　　Stephens's Satyr
　　　(W. G. Wright)

　6. *Cercyonis sthenele*　　Sthenele Satyr
　　　(Boisduval)
　 †a. *sthenele* (Boisduval)
　　b. *silvestris*　　　　　Woodland Satyr
　　　(W. H. Edwards)
　　c. *paula* (W. H. Edwards)　Little Satyr

　7. *Cercyonis oeta* (Boisduval)　Least Satyr

8. *Oeneis nevadensis*   Great Arctic
   (C. & R. Felder)
   a. *nevadensis*
      (C. & R. Felder)
   b. *iduna* (W. H. Edwards)   Iduna Arctic

9. *Oeneis ivallda* (Mead)   Ivallda Arctic

*10. *Oeneis chryxus*   Chryxus Arctic
     (Doubleday & Hewitson)
     a. *stanislaus* Hovanitz   Chryxus Arctic

Family DANAIDAE   Milkweed Butterflies

11. *Danaus plexippus*   Monarch
    (Linnaeus)

*12. *Danaus gilippus* (Cramer)   Queen
     a. *strigosus* (Bates)   Striated Queen

Family HELICONIIDAE   Long-wings

*13. *Agraulis vanillae*   Gulf Fritillary
     (Linnaeus)
     a. *incarnata* (Riley)   Gulf Fritillary

Family NYMPHALIDAE   Brush-footed Butterflies

14. *Euptoieta claudia* (Cramer)   Variegated Fritillary

*15. *Speyeria cybele* (Fabricius) Great Spangled Fritillary
     a. *leto* (Behr)   Leto Fritillary

*16. *Speyeria nokomis*   Nokomis Fritillary
     (W. H. Edwards)
     a. *apacheana* (Skinner)   Apache Fritillary

17. *Speyeria coronis* (Behr)   Crown Fritillary
    a. *coronis* (Behr)
    b. *hennei* (Gunder)   Henne's Fritillary
    c. *semiramis*   Semiramis Fritillary
       (W. H. Edwards)
    d. nr. *simaetha* dos Passos   Simaetha Fritillary
       & Gray
    e. *snyderi* (Skinner)   Snyder's Fritillary

18. *Speyeria zerene*   Zerene Fritillary
    (Boisduval)
    a. *zerene* (Boisduval)
    b. *conchyliatus*   Royal Fritillary
       (J. A. Comstock)
    c. *malcolmi*   Malcolm's Fritillary
       (J. A. Comstock)

    d. *behrensii*   Behrens's Fritillary
       (W. H. Edwards)

    e. *myrtleae* dos Passos &   Myrtle's Fritillary
       Grey

    f. *gunderi*   Gunder's Fritillary
       (J. A. Comstock)
       (= *cynna* dos Passos &
       Grey)

    g. *gloriosa* Moeck   Glorious Fritillary

19. *Speyeria callippe*   Callippe Fritillary
    (Boisduval)

    a. *callippe* (Boisduval)

    b. *liliana* (Hy. Edwards)   Lilian's Fritillary

    c. *rupestris* (Behr)   Rupestris Fritillary

    d. *juba* (Boisduval)   Yuba Fritillary

    e. *sierra* dos Passos & Grey   Sierra Fritillary

    f. *nevadensis*   Nevada Fritillary
       (W. H. Edwards)

    g. *comstocki* (Gunder)   Comstock's Fritillary

    h. *macaria* (W. H. Edwards)   Macaria Fritillary
       form "*laurina*"   Laurina Fritillary
       (W. G. Wright)

    i. *inornata*   Plain Fritillary
       (W. H. Edwards)

20. *Speyeria adiaste*   Unsilvered Fritillary
    (W. H. Edwards)

    a. *adiaste* (W. H. Edwards)

    b. *clemencei*   Clemence's Fritillary
       (J. A. Comstock)

 †c. *atossa* (W. H. Edwards)   Atossa Fritillary

21. *Speyeria egleis* (Behr)   Egleis Fritillary

    a. *egleis* (Behr)

    b. *oweni* (W. H. Edwards)   Owen's Fritillary

    c. *tehachapina*   Tehachapi Fritillary
       (J. A. Comstock)

*22. *Speyeria atlantis*   Atlantis Fritillary
    (W. H. Edwards)

    a. *irene* (Boisduval)   Irene Fritillary

    b. *dodgei* (Gunder)   Dodge's Fritillary

23. *Speyeria hydaspe*   Hydaspe Fritillary
    (Boisduval)

    a. *hydaspe* (Boisduval)

    b. *purpurascens*   Purple Fritillary
       (Hy. Edwards)

c. *viridicornis* (J. A. Comstock)	Greenhorn Fritillary
*24. *Speyeria mormonia* (Boisduval)	Mormon Fritillary
a. *arge* (Strecker)	Arge Fritillary
b. *erinna* (W. H. Edwards)	Erinna Fritillary
25. *Clossiana epithore* (W. H. Edwards)	Western Meadow Fritillary
a. *epithore* (W. H. Edwards)	
b. *chermocki* (E. & S. Perkins)	Chermock's Meadow Fritillary
c. *sierra* (E. Perkins & Meyer)	Sierra Meadow Fritillary
26. *Occidryas chalcedona* (Doubleday)	Common Checkerspot
a. *chalcedona* (Doubleday)	
b. *dwinellei* (Hy. Edwards)	Dwinelle's Checkerspot
c. *macglashanii* (Rivers)	McGlashan's Checkerspot
d. *sierra* (W. G. Wright)	Sierra Checkerspot
e. *olancha* (W. G. Wright)	Olancha Checkerspot
f. *hennei* (Scott) (*quino* of authors, not Behr)	Henne's Checkerspot
g. *kingstonensis* (T. & J. Emmel)	Kingston Checkerspot
h. *corralensis* (T. & J. Emmel)	Corral Checkerspot
27. *Occidryas colon* (W. H. Edwards)	Colon Checkerspot
28. *Occidryas editha* (Boisduval)	Editha Checkerspot
a. *editha* (Boisduval)	
b. *bayensis* (Sternitsky)	Bay Region Checkerspot
c. *luestherae* (Murphy & Ehrlich)	Luesther's Checkerspot
d. *monoensis* (Gunder)	Mono Checkerspot
e. *rubicunda* (Hy. Edwards)	Ruddy Checkerspot
f. *nubigena* (Behr)	Cloud-born Checkerspot
g. *aurilacus* (Gunder)	Gold Lake Checkerspot
h. *baroni* (W. H. Edwards)	Baron's Checkerspot
i. *edithana* (Strand)	Strand's Checkerspot
j. *augustina* (Scott) (*augusta* of authors, not W. H. Edwards)	Augustina Checkerspot

k. *quino* (Behr)   Quino Checkerspot
  [*wrighti* (Gunder)]
  [*augusta*
  (W. H. Edwards)]
l. *insularis* (T. & J. Emmel) Island Checkerspot

29. *Charidryas gabbii* (Behr) Gabb's Checkerspot

30. *Charidryas neumoegeni* Neumoegen's Checkerspot
  (Skinner)

31. *Charidryas acastus*   Acastus Checkerspot
  (W. H. Edwards)

32. *Charidryas palla*   Northern Checkerspot
  (Boisduval)
 a. *palla* (Boisduval)
 b. *whitneyi* (Behr)   Whitney's Checkerspot
 c. *vallismortis*   Death Valley Checkerspot
  (J. W. Johnson)

*33. *Charidryas damoetas* Damoetas Checkerspot
  (Skinner)
 a. *malcolmi*   Malcolm's Checkerspot
  (J. A. Comstock)

34. *Charidryas hoffmanni* Hoffmann's Checkerspot
  (Behr)
 a. *hoffmanni* (Skinner)
 b. *segregata* (Barnes & Segregated Checkerspot
  McDunnough)

*35. *Chlosyne lacinia* (Geyer) Bordered Patch
 a. *crocale* (W. H. Edwards) Crocale Patch
  form "*rufescens*" Rufescent Patch
  (W. H. Edwards)
  form "*nigrescens*" Dusky Patch
  (W. H. Edwards)

36. *Chlosyne californica* California Patch
  (W. G. Wright)

37. *Thessalia leanira*   Leanira Checkerspot
  (C. & R. Felder)
 a. *leanira* (C. & R. Felder)
 b. *daviesi* (Wind)   Davies's Checkerspot
 c. *wrighti* (W. H. Edwards) Wright's Checkerspot
 d. *cerrita* (W. G. Wright) Cerrita Checkerspot
 e. *alma* (Strecker)   Alma Checkerspot

*38. *Dymasia chara*   Chara Checkerspot
  (W. H. Edwards)
 a. *imperialis* (Bauer) Imperial Checkerspot

*39. *Poladryas arachne*   Arachne Checkerspot
    (W. H. Edwards)
    a. *monache*   Monache Checkerspot
      (J. A. Comstock)

*40. *Phyciodes pascoensis*   Northern Pearl Crescent
    (W. G. Wright)
    a. *distinctus* (Bauer)   Distinct Crescent

41. *Phyciodes phaon*   Phaon Crescent
    (W. H. Edwards)

42. *Phyciodes campestris* Behr)   Field Crescent
    a. *campestris* (Behr)
    b. *montanus* (Behr)   Mountain Crescent

43. *Phyciodes mylitta*   Mylitta Crescent
    (W. H. Edwards)

44. *Phyciodes orseis*   Orseis Crescent
    W. H. Edwards
    a. *orseis* W. H. Edwards
    b. *herlani* Bauer   Herlan's Crescent

*45. *Polygonia satyrus*   Satyr Anglewing
    (W. H. Edwards)
    a. *neomarsyas* dos Passos   Satyr Anglewing

*46. *Polygonia faunus*   Green Comma
    (W. H. Edwards)
    a. *rusticus* (W. H. Edwards)   Rustic Anglewing

47. *Polygonia oreas*   Oreas Anglewing
    (W. H. Edwards)

48. *Polygonia silvius*   Sylvan Anglewing
    (W. H. Edwards)

49. *Polygonia zephyrus*   Zephyr Anglewing
    (W. H. Edwards)

50. *Nymphalis californica*   California Tortoise Shell
    (Boisduval)

51. *Nymphalis antiopa*   Mourning Cloak
    (Linnaeus)

*52. *Aglais milberti* (Godart)   Milbert's Tortoise Shell
    a. *furcillata* (Say)   Milbert's Tortoise Shell

*53. *Vanessa atalanta* (Linnaeus)   Red Admiral
    a. *rubria* (Fruhstorfer)   Red Admiral

54. *Vanessa cardui* (Linnaeus)   Painted Lady

55. *Vanessa virginiensis* (Drury)   American Painted Lady

56. *Vanessa annabella* (Field) West Coast Lady

57. *Junonia coenia* Hübner Buckeye

58. *Junonia nigrosuffusa* Barnes Dark Buckeye
 & McDunnough

*59. *Basilarchia archippus* Viceroy
 (Cramer)
 a. *obsoleta* Arizona Viceroy
  (W. H. Edwards)

*60. *Basilarchia weidemeyerii* Weidemeyer's Admiral
 (W. H. Edwards)
 a. *latifascia* Wide-banded Admiral
  (E. & S. Perkins)
  hybrid *latifascia* x Friday's Admiral
  lorquini

61. *Basilarchia lorquini* Lorquin's Admiral
 (Boisduval)

*62. *Adelpha bredowii* Geyer Sister
 a. *californica* (Butler) California Sister
 b. *eulalia* (Doubleday & Arizona Sister
  Hewitson)

Family PAPILIONIDAE Swallowtails and Parnassians

63. *Parnassius clodius* Clodius Parnassian
 Ménétriés
 a. *clodius* Ménétriés
 b. *sol* Bryk & Eisner Sol Parnassian
 †c. *strohbeeni* Sternitzky Strohbeen's Parnassian
 d. *baldur* W. H. Edwards Baldur Parnassian

*64. *Parnassius phoebus* Small Apollo
 (Fabricius)
 a. *sternitzkyi* McDunnough Sternitzky's Parnassian
 b. *behrii* W. H. Edwards Behr's Parnassian

65. *Battus philenor* (Linnaeus) Pipe-vine Swallowtail
 a. *philenor* (Linnaeus)
 b. *hirsutus* (Skinner) Hairy Pipe-vine Swallowtail

66. *Papilio bairdii* Baird's Swallowtail
 W. H. Edwards
  form "*brucei*" Bruce's Swallowtail
  W. H. Edwards
  form "*hollandi*" Holland's Swallowtail
  W. H. Edwards

67. *Papilio zelicaon* Lucas Anise Swallowtail

*68. *Papilio polyxenes* Fabricius   Polyxenes Swallowtail
     a. *coloro* W. G. Wright   Wright's Swallowtail
       (*rudkini*   (Rudkin's Swallowtail)
       F. & R. Chermock)
       form "*clarki*"   Clark's Swallowtail
       F. & R. Chermock
       form "*comstocki*"   Comstock's Swallowtail
       F. & R. Chermock

69. *Papilio indra* Reakirt   Indra Swallowtail
     a. *indra* Reakirt   Short-tailed Swallowtail
     b. *pergamus* Hy. Edwards   Edwards's Swallowtail
     c. *fordi* J. A. Comstock &   Ford's Swallowtail
       Martin
     d. *martini* T. & J. Emmel   Martin's Swallowtail
     e. *phyllisae* J. Emmel   Phyllis's Swallowtail
     f. *panamintensis* J. Emmel   Panamint Swallowtail

70. *Papilio cresphontes* Cramer   Giant Swallowtail

71. *Papilio rutulus* Lucas   Western Tiger Swallowtail

72. *Papilio eurymedon* Lucas   Pale Swallowtail

73. *Papilio multicaudatus*   Two-tailed Swallowtail
     W. F. Kirby

Family PIERIDAE   Whites, Sulfurs, Marbles, and Orange-tips

74. *Neophasia menapia*   Pine White
     (C. & R. Felder)
     a. *menapia*
       (C. & R. Felder)
     b. *melanica* Scott   Coastal Pine White

75. *Pontia beckerii*   Becker's White
     (W. H. Edwards)

76. *Pontia sisymbrii* (Boisduval)   California White

77. *Pontia protodice* (Boisduval   Common White
     & Le Conte)

78. *Pontia occidentalis* (Reakirt)   Western White
     a. *occidentalis* (Reakirt)
       form "*calyce*"   Calyce White
       (W. H. Edwards)

*79. *Artogeia napi* (Linnaeus)   Mustard White
     a. *venosa* (Scudder)   Veined White
     b. *microstriata*   Small-veined White
       (J. A. Comstock)

c. *marginalis* (Scudder) form "*pallida*" (Scudder)	Margined White Pallid White
80. *Artogeia rapae* (Linnaeus)	Cabbage Butterfly
81. *Colias eurytheme* Boisduval form "*amphidusa*" Boisduval	Alfalfa Butterfly Flavid Sulfur
*82. *Colias philodice* Godart a. *eriphyle* W. H. Edwards	Clouded Sulfur Clouded Sulfur
*83. *Colias occidentalis* Scudder a. *chrysomelas* Hy. Edwards	Western Sulfur Black and Gold Sulfur
84. *Colias harfordii* Hy. Edwards	Harford's Sulfur
*85. *Colias alexandra* W. H. Edwards a. *edwardsi* W. H. Edwards	Alexandra Sulfur Edwards's Sulfur
86. *Colias behrii* W. H. Edwards	Behr's Sulfur
87. *Zerene eurydice* (Boisduval) form "*bernardino*" (W. H. Edwards) form "*amorphae*" (Hy. Edwards)	California Dog-face
88. *Zerene cesonia* (Stoll)	Southern Dog-face
*89. *Phoebis sennae* (Linnaeus) a. *marcellina* (Cramer)	Cloudless Sulfur Cloudless Sulfur
90. *Phoebis agarithe* (Boisduval)	Large Orange Sulfur
91. *Eurema mexicana* (Boisduval)	Mexican Yellow
92. *Eurema nicippe* (Cramer)	Nicippe Yellow
93. *Nathalis iole* Boisduval	Dainty Sulfur
94. *Anthocharis cethura* (C. & R. Felder) a. *cethura* (C. & R. Felder) b. *morrisoni* W. H. Edwards c. *caliente* W. G. Wright d. *deserti* W. G. Wright e. *catalina* Meadows f. *pima* W. H. Edwards	Felders' Orange-tip Morrison's Orange-tip Caliente Orange-tip Desert Orange-tip Catalina Orange-tip Pima Orange-tip

95. *Anthocharis sara* Lucas   Sara Orange-tip
    a. *sara* Lucas
      form "*reakirtii*"   Reakirt's Orange-tip
      W. H. Edwards
    b. *stella* W. H. Edwards   Stella Orange-tip
    c. *thoosa* (Scudder)   Thoosa Orange-tip
    d. *gunderi* Ingham   Gunder's Orange-tip

96. *Falcapica lanceolata*   Boisduval's Marble
    (Lucas)
    a. *lanceolata* (Lucas)
    b. *australis* (F. Grinnell)   Grinnell's Marble

97. *Euchloe hyantis*   Edwards's Marble
    (W. H. Edwards)
    a. *hyantis* (W. H. Edwards)
    b. *lotta* (Beutenmüller)   Southern Marble
    c. *andrewsi* Martin   Martin's Marble

98. Euchloe ausonides (Lucas)   Large Marble

Family LIBYTHEIDAE   Snout Butterflies

*99. *Libytheana bachmanii*   Snout Butterfly
    (Kirtland)
    a. *larvata* (Strecker)   Snout Butterfly

Family RIODINIDAE   Metalmarks

100. *Apodemia mormo*   Mormon Metalmark
    (C. & R. Felder)
    a. *mormo* (C. & R. Felder)
    b. *langei* J. A. Comstock   Lange's Metalmark
    c. *tuolumnensis* Opler &   Tuolumne Metalmark
      Powell
    d. *cythera* (W. H. Edwards)   Cythera Metalmark
    e. *virgulti* (Behr)   Behr's Metalmark
    f. *deserti* Barnes &   Desert Metalmark
      McDunnough
    g. nr. *dialeuca* Opler &   Whitish Metalmark
      Powell

*101. *Apodemia palmerii*   Palmer's Metalmark
    (W. H. Edwards)
    a. *marginalis* (Skinner)   Palmer's Metalmark

*102. *Calephelis nemesis*   Fatal Metalmark
    (W. H. Edwards)
    a. *californica* McAlpine   Dusky Metalmark
    b. *dammersi* McAlpine   Dammers's Metalmark

103. *Calephelis wrighti* Holland   Wright's Metalmark

Family LYCAENIDAE

Hairstreaks, Coppers, and Blues

Subfamily THECLINAE

104. *Habrodais grunus* (Boisduval)
    a. *grunus* (Boisduval)
    b. *lorquini* Field
    c. *herri* Field

Boisduval's Hairstreak

Lorquin's Hairstreak
Herr's Hairstreak

Subfamily EUMAEINAE

*105. *Chlorostrymon simaethis* (Drury)
    a. *sarita* (Skinner)

Simaethis Hairstreak

Sarita Hairstreak

*106. *Harkenclenus titus* (Fabricius)
    a. *immaculosus* (W. P. Comstock)

Coral Hairstreak

Coral Hairstreak

107. *Satyrium fuliginosum* (W. H. Edwards)

Sooty Gossamer-wing

108. *Satyrium behrii* (W. H. Edwards)

Behr's Hairstreak

109. *Satyrium auretorum* (Boisduval)
    a. *auretorum* (Boisduval)
    b. *spadix* (Hy. Edwards)

Gold-hunter's Hairstreak

Nut-brown Hairstreak

110. *Satyrium tetra* (W. H. Edwards)

Gray Hairstreak

111. *Satyrium saepium* (Boisduval)
    a. *saepium* (Boisduval)
    b. *fulvescens* (Hy. Edwards)
    c. *chlorophora* (Watson & W. P. Comstock)
    d. *chalcis* (W. H. Edwards)

Hedge-row Hairstreak

Tawny Hairstreak
Purplish Hairstreak

Bronzed Hairstreak

112. *Satyrium sylvinum* (Boisduval)
    a. *sylvinum* (Boisduval)
    b. *desertorum* (F. Grinnell)

Sylvan Hairstreak

Desert Hairstreak

113. *Satyrium dryope* (W. H. Edwards)

Dryope Hairstreak

114. *Satyrium californicum* (W. H. Edwards)

California Hairstreak

115. *Ministrymon leda* Leda Hairstreak
(W. H. Edwards)
form "*ines*" Ines Hairstreak
(W. H. Edwards)

116. *Mitoura loki* (Skinner) Skinner's Hairstreak

117. *Mitoura thornei* J. W. Brown Thorne's Hairstreak

118. *Mitoura siva* Siva Hairstreak
(W. H. Edwards)
a. *siva* (W. H. Edwards)
b. *juniperaria* Juniper Hairstreak
J. A. Comstock
c. *mansfieldi* Tilden Mansfield's Hairstreak
d. *chalcosiva* Clench Clench's Hairstreak

119. *Mitoura barryi* K. Johnson Barry's Hairstreak

120. *Mitoura nelsoni* (Boisduval) Nelson's Hairstreak

121. *Mitoura muiri* Muir's Hairstreak
(Hy. Edwards)

122. *Mitoura spinetorum* Thicket Hairstreak
(Hewitson)

123. *Mitoura johnsoni* (Skinner) Johnson's Hairstreak

*124. *Incisalia mossii* Moss's Hairstreak
(Hy. Edwards)
a. *bayensis* (R. M. Brown) San Bruno Elfin
b. *windi* Clench Wind's Hairstreak
c. *doudoroffi* dos Passos Doudoroff's Hairstreak

125. *Incisalia fotis* (Strecker) Fotis Hairstreak

*126. *Incisalia augusta* (W. Kirby) Brown Elfin
a. *iroides* (Boisduval) Western Brown Elfin
b. nr. *annetteae* dos Passos Annette's Brown Elfin

127. *Incisalia eryphon* Western Banded Elfin
(Boisduval)

128. *Callophrys dumetorum* Bramble Hairstreak
(Boisduval)
a. *dumetorum* (Boisduval)
b. *perplexa* Barnes & Southern Bramble Hairstreak
Benjamin

129. *Callophrys viridis* Green Hairstreak
(W. H. Edwards)

130. *Callophrys comstocki* Henne Comstock's Hairstreak

131. *Callophrys lemberti* Tilden Lembert's Hairstreak

*132. *Atlides halesus* (Cramer) Great Purple Hairstreak
a. *estesi* Clench Great Purple Hairstreak

133. *Strymon avalona*   Avalon Hairstreak
    (W. G. Wright)

*134. *Strymon melinus* Hübner   Common Hairstreak
    a. *pudicus* (Hy. Edwards)   Common Hairstreak

*135. *Strymon columella*   Columella Hairstreak
    (Fabricius)
    a. *istapa* (Reakirt)   Columella Hairstreak

Subfamily LYCAENINAE

136. *Tharsalea arota* (Boisduval)   Tailed Copper
    a. *arota* (Boisduval)   Arota Copper
    b. *nubila* J. A. Comstock   Cloudy Copper
    c. *virginiensis*   Virginia Copper
       (W. H. Edwards)

*137. *Lycaena phlaeas* (Linnaeus)   Small Copper
    a. *hypophlaeas* (Boisduval)   American Copper

138. *Lycaena cuprea*   Lustrous Copper
    (W. H. Edwards)

139. *Gaeides xanthoides*   Great Copper
    (Boisduval)
    a. *xanthoides* (Boisduval)
    b. *luctuosa* (Watson &   Mourning-garbed Copper
       W. P. Comstock)

140. *Gaeides editha* (Mead)   Edith's Copper

141. *Gaeides gorgon* (Boisduval)   Gorgon Cooper

142. *Chalceria rubida* (Behr)   Ruddy Cooper
    a. *rubida* (Behr)
    b. *monachensis* (K. Johnson   Monache Copper
       & Balogh)

143. *Chalceria heteronea*   Varied Blue
    (Boisduval)
    a. *heteronea* (Boisduval)
    b. *clara* (Hy. Edwards)   Bright Blue Copper

144. *Epidemia helloides*   Purplish Copper
    (Boisduval)

145. *Epidemia nivalis*   Nivalis Copper
    (Boisduval)

146. *Epidemia mariposa*   Mariposa Copper
    (Reakirt)

147. *Hermelycaena hermes*   Hermes Copper
    (W. H. Edwards)

Subfamily POLYOMMATINAE

148. *Brephidium exile*   Pygmy Blue
     (Boisduval)

149. *Leptotes marina* (Reakirt)   Marine Blue

*150. *Hemiargus ceraunus*   Ceraunus Blue
      (Fabricius)
      a. *gyas* (W. H. Edwards)   Edwards's Blue

*151. *Hemiargus isola* (Reakirt)   Reakirt's Blue
      a. *alce* (W. H. Edwards)   Reakirt's Blue

*152. *Lycaeides idas* (Linnaeus)   Northern Blue
      a. *lotis* (Lintner)   Lotis Blue
      b. *anna* (W. H. Edwards)   Anna Blue
      c. *ricei* (Cross)   Rice's Blue

*153. *Lycaeides melissa*   Melissa Blue
      (W. H. Edwards)
      a. *paradoxa*   Orange-margined Blue
         (F. H. Chermock)
      b. *fridayi* (F. H. Chermock)   Friday's Blue
      (c. unnamed population)

154. *Plebejus saepiolus*   Greenish Blue
     (Boisduval)
     a. *saepiolus* (Boisduval)
     b. *hilda* (J. & F. Grinnell)   Hilda Blue

155. *Plebulina emigdionis*   San Emigdio Blue
     (F. Grinnell)

156. *Icaricia icarioides*   Icarioides Blue
     (Boisduval)
     a. *icarioides* (Boisduval)
     b. *fulla* (W. H. Edwards)   Fulla Blue
     c. *mintha* W. H. Edwards)   Mintha Blue
     d. *helios* (W. H. Edwards)   Helios Blue
     e. *evius* (Boisduval)   Evius Blue
     f. *moroensis* (Sternitzky)   Morro Blue
     g. *missionensis* (Hovanitz)   Mission Blue
     h. *ardea* (W. H. Edwards)   Ardea Blue
     i. *pardalis* (Behr)   Pardalis Blue
     †j. *pheres* (Boisduval)   Pheres Blue

157. *Icaricia shasta*   Shasta Blue
     (W. H. Edwards)

158. *Icaricia acmon* (Westwood   Acmon Blue
     & Hewitson)
     a. *acmon* (Westwood &
        Hewitson)

form "*cottlei*" (F. Grinnell)	Cottle's Blue
b. nr. *texana* (Goodpasture)	Texas Blue
159. *Icaricia lupini* (Boisduval)	Lupine Blue
a. *lupini* (Boisduval)	
b. *monticola* (Clemence)	Clemence's Blue
c. *chlorina* (Skinner)	Skinner's Blue
160. *Icaricia neurona* (Skinner)	Veined Blue
*161. *Agriades franklinii* (Curtis)	Arctic Blue
a. *podarce* (C. & R. Felder)	Gray Blue
162. *Everes comyntas* (Godart)	Eastern Tailed Blue
163. *Everes amyntula* (Boisduval)	Western Tailed Blue
164. *Euphilotes battoides* (Behr)	Square-spotted Blue
a. *battoides* (Behr)	
b. *intermedia* (Barnes & McDunnough)	Intermediate Blue
c. *bernardino* (Barnes & McDunnough)	San Bernardino Blue
d. *glaucon* (W. H. Edwards)	Glaucous Blue
e. *martini* (Mattoni)	Martin's Blue
f. *allyni* (Shields)	El Segundo Blue
g. *baueri* (Shields)	Bauer's Blue
h. *comstocki* (Shields)	Comstock's Blue
165. *Euphilotes enoptes* (Boisduval)	Dotted Blue
a. *enoptes* (Boisduval)	
b. *bayensis* (Langston)	Bay Region Blue
c. *tildeni* (Langston)	Tilden's Blue
d. *dammersi* (J. A. Comstock & Henne)	Dammers's Blue
e. *smithi* (Mattoni)	Smith's Blue
f. *langstoni* (Shields)	Langston's Blue
166. *Euphilotes mojave* (Watson & W. H. Comstock)	Mojave Blue
*167. *Euphilotes pallescens* (Tilden & Downey)	Pallid Blue
a. *elvirae* (Mattoni)	Elvira's Blue
168. *Philotiella speciosa* (Hy. Edwards)	Small Blue
a. *speciosa* (Hy. Edwards)	
b. *bohartorum* (Tilden)	Bohart's Blue

169. *Philotes sonorensis*   Sonoran Blue
(C. & R. Felder)

*170. *Glaucopsyche lygdamus*   Silvery Blue
(Doubleday)
   a. *incognitus* Tilden   Behr's Blue
   b. *columbia* (Skinner)   Columbia Blue
   c. *australis* (F. Grinnell) Southern Blue
   d. *palosverdesensis*   Palos Verdes Blue
      (E. Perkins & J. Emmel)

171. *Glaucopsyche piasus*   Arrowhead blue
(Boisduyal)
   a. *piasus* (Boisduval)
   b. *catalina* (Reakirt)   Coastal Arrowhead Blue

†172. *Glaucopsyche xerces*   Xerces Blue
(Boisduval)
   form "*polyphemus*"   Polyphemus Blue
   (Boisduval)

*173. *Celastrina ladon* (Cramer) Spring Azure
   a. *echo* (W. H. Edwards)   Echo Blue
   b. *cinerea* (W. H. Edwards) Cinereous Blue

Family MEGATHYMIDAE   Giant Skippers

174. *Agathymus stephensi*   Stephens's Giant Skipper
(Skinner)

175. *Agathymus alliae*   Allie's Giant Skipper
(D. Stallings & Turner)

176. *Agathymus baueri*   Bauer's Giant Skipper
(D. Stallings & Turner)

*177. *Megathymus coloradensis* Common Giant Skipper
C. V. Riley
   a. *martini* D. Stallings & Martin's Giant Skipper
     Turner
   b. *maudae* D. Stallings,   Maud's Giant Skipper
     Turner, & J. Stallings

Family HESPERIIDAE   Skippers
Subfamily HESPERIINAE

178. *Panoquina errans* (Skinner) Wandering Skipper

179. *Calpodes ethlius* (Stoll)   Brazilian Skipper

180. *Nyctelius nyctelius*   Nyctelius Skipper
(Latreille)

181. *Lerodea eufala*   Eufala Skipper
(W. H. Edwards)

182. *Amblyscirtes vialis*
(W. H. Edwards)   Roadside Skipper

183. *Euphyes ruricola*
(Boisduval)   Dun Skipper

184. *Paratrytone melane*
(W. H. Edwards)   Umber Skipper

185. *Ochlodes sylvanoides*
(Boisduval)   Woodland Skipper
   a. *sylvanoides* (Boisduval)
   b. *santacruza* Scott   Santa Cruz Skipper

186. *Ochlodes pratincola*
(Boisduval)   Meadow Skipper

187. *Ochlodes agricola*
(Boisduval)   The Farmer
   a. *agricola* (Boisduval)
   b. *verus* (W. H. Edwards)   The Verus Farmer

188. *Ochlodes yuma*
(W. H. Edwards)   Yuma Skipper

189. *Atalopedes campestris*
(Boisduval)   Field Skipper

190. *Polites sabuleti* (Boisduval)   Sandhill Skipper
   a. *sabuleti* (Boisduval)
   b. *tecumseh* (F. Grinnell)   Tecumseh Skipper
   c. *chusca* (W. H. Edwards)   Chusca Skipper

191. *Polites themistocles*
(Latreille)   Tawny-edged Skipper

192. *Polites sonora* (Scudder)   Sonora Skipper
   a. *sonora* (Scudder)
   b. *siris* (W. H. Edwards)   Dog-star Skipper

*193. *Hesperia uncas*
W. H. Edwards   Uncas Skipper
   a. *macswaini* MacNeill   Uncas Skipper

*194. *Hesperia comma* (Linnaeus)   Comma Skipper
   a. *harpalus*   Harpalus Skipper
     (W. H. Edwards)
   b. *yosemite* Leussler   Yosemite Skipper
   c. *oregonia*   Oregon Skipper
     (W. H. Edwards)
   d. *dodgei* (Bell)   Dodge's Skipper
   e. *tildeni* H. A. Freeman   Tilden's Skipper
   f. *leussleri* Lindsey   Leussler's Skipper

195. *Hesperia nevada* (Scudder)   Nevada Skipper

196. *Hesperia miriamae*   Miriam's Skipper
MacNeill

197. *Hesperia lindseyi* (Holland)   Lindsey's Skipper

198. *Hesperia columbia*   Columbian Skipper
(Scudder)

*199. *Hesperia pahaska* (Leussler) Pahaska Skipper
a. *martini* MacNeill   Pahaska Skipper

200. *Hesperia juba* (Scudder)   Yuba Skipper

201. *Pseudocopaeodes eunus*   Eunus Skipper
(W. H. Edwards)

202. *Yvretta carus*   Carus Skipper
(W. H. Edwards)

203. *Hylephila phyleus* (Drury) Fiery Skipper

204. *Copaeodes aurantiaca*   Hewitson's Skipper
(Hewitson)

205. *Nastra julia*   Julia's Skipper
(H. A. Freeman)

206. *Nastra neamathla* (Skinner Neamathla Skipper
& R. C. Williams)

Subfamily HETEROPTERINAE

207. *Piruna pirus*   Pirus Skipperling
(W. H. Edwards)

*208. *Carterocephalus palaemon* Arctic Skipper
(Pallas)
a. *mandan* (W. H. Edwards) Arctic Skipper

Subfamily PYRGINAE

209. *Pholisora catullus*   Common Sooty-wing
(Fabricius)

210. *Pholisora libya* (Scudder) Mojave Sooty-wing

*211. *Pholisora alpheus*   Alpheus Sooty-wing
(W. H. Edwards)
a. *oricus* W. H. Edwards   Alpheus Sooty-wing

212. *Pholisora gracielae*   MacNeill's Sooty-wing
MacNeill

213. *Heliopetes domicella*   Erichson's Skipper
(Erichson)

214. *Heliopetes ericetorum*   Large White Skipper
(Boisduval)

215. *Heliopetes laviana*   Laviana Skipper
(Hewitson)

216. *Pyrgus scriptura* (Boisduval) Little Checkered Skipper

217. *Pyrgus ruralis* (Boisduval) Rural Skipper
  a. *ruralis* (Boisduval)
  b. *lagunae* Scott Laguna Mountains Rural Skipper

218. *Pyrgus communis* (Grote) Common Checkered Skipper

219. *Pyrgus albescens* Plötz Western Checkered Skipper

220. *Erynnis icelus* (Scudder & Burgess) Dreamy Dusky-wing

*221. *Erynnis brizo* (Boisduval & Le Conte) Sleepy Dusky-wing
  a. *lacustra* (W. G. Wright) Wright's Dusky-wing
  b. *burgessi* (Skinner) Burgess's Dusky-wing

222. *Erynnis persius* (Scudder) Persius Dusky-wing

223. *Erynnis afranius* (Lintner) Afranius Dusky-wing

224. *Erynnis funeralis* (Scudder & Burgess) Funereal Dusky-wing

*225. *Erynnis pacuvius* (Lintner) Pacuvius Dusky-wing
  a. *callidus* (F. Grinnell) Artful Dusky-wing
  b. *pernigra* (F. Grinnell) Grinnell's Dusky-wing
  c. *lilius* (Dyar) Dyar's Dusky-wing

226. *Erynnis tristis* (Boisduval) Mournful Dusky-wing

227. *Erynnis propertius* (Scudder & Burgess) Propertius Dusky-wing

228. *Systasea zampa* (W. H. Edwards) Powdered Skipper

229. *Staphylus ceos* (W. H. Edwards) Ceos Sooty-wing

230. *Thorybes pylades* (Scudder) Northern Cloudy-wing

231. *Thorybes diversus* Bell Diverse Cloudy-wing

*232. *Thorybes mexicana* (Herrich-Schäffer) Mexican Cloudy-wing
  a. *nevada* Scudder Nevada Cloudy-wing
  b. *aemilea* (Skinner) Emily's Cloudy-wing
  c. *blanca* Scott White Mountains Cloudy-wing

233. *Urbanus proteus* (Linnaeus) Long-tailed Skipper

234. *Urbanus simplicius* (Stoll) Simplicius Skipper

*235. *Polygonus leo* (Gmelin) Hammock Skipper
  a. *arizonensis* (Skinner) Arizona Skipper

236. *Epargyreus clarus* (Cramer) Silver-spotted Skipper
 a. *clarus* (Cramer)
 b. *californicus* (J. B. Smith) California Silver-spotted Skipper
 c. *huachuca* Dixon Arizona Silver-spotted Skipper

GLOSSARY

Boldfaced terms are defined within the glossary.

abdomen Third (last) section of an insect body. Contains digestive and reproductive organs; bears no legs. (See Figure 25.)

aberrant Of abnormal appearance (sometimes called "freakish").

androconia Scent scales, also called sex scales, found in localized units in certain places on male insects. Example: the **stigma** of a male skipper.

antenna(ae) Paired sensory organs arising near the eyes on the head; the feelers. (See Figures 23, 25.)

anus Posterior opening of the digestive tract; the vent.

apiculus The sharp-pointed outer end of the antennal club in Hesperiidae.

appressed Closely applied to, as **palpi** against the face of a butterfly.

arthropod An organism whose body is made up of rings (segments) to which jointed appendages are attached. Includes crabs, spiders, insects, and others.

aurora A row of red or orange spots, sometimes confluent, along the wing edge (usually the hind wing) of butterflies such as blues.

bilobed Divided into two lobes.

boreal From or belonging to the north; said of organisms living in northern regions.

brood All offspring hatching at one time from one series of parents, usually maturing at nearly the same time.

caterpillar The **larva** or second stage of a butterfly or moth (Lepidoptera), a sawfly (Hymenoptera), or a scorpionfly (Mecoptera). (See Figure 3.)

caudal As used here, of or pertaining to the posterior end of the body.

cell bar A marking, usually dark, across the cell of a butterfly wing.

cell spot A spot marking, usually dark, in the cell of a butterfly wing.

chaparral A type of vegetation characterized by rigid brush; "elfin forest"; brushland.

character A quality of form, color, or structure by which an organism may be recognized or differentiated from others.

chlorocresol A crystalline chemical used by some insect collectors for storing specimens prior to mounting; tends to keep the specimens relaxed.

chrysalis The naked **pupa** of a butterfly; the reorganizational or "resting" stage that follows the **larva** (**caterpillar**).

cismontane "On this side of the mountain"; in northern California, said of the area west of the Sierra Nevada–Cascade crest; in southern California, said of the area south of the Transverse Ranges and west of the Peninsular Ranges.

clasper The valve or **harpe** of the male genitalia of **Lepidoptera**; or, as used by some **lepidopterists**, a specialized clasping organ on the inner face of the harpe or valve.

cocoon A silken covering of the **pupa**, spun by the **larva** prior to **pupation**.

copulation Sexual intercourse; mating.

costa The upper edge of an insect wing. (See Figure 25.)

costal bar A dark, bar-shaped marking, extending from the **costal** margin of the forewing a short distance inward.

costal fold A rolled edge on the forewing **costa** of some skippers.

cremaster Small hooks at the posterior end of a **pupa**, by which it is suspended.

crepuscular Active at twilight or just before dawn.

crochets The spines or hooks on the larval **prolegs**, and the similar structures at the tip of the pupal **abdomen**, by which the **pupa** (**chrysalis**) is suspended.

cryptic Hidden, concealed; said of organisms that blend into their surroundings and so are difficult to see.

Cubitus 2 vein Cu_2, second cubital vein. (See Figure 25.)

danaid A butterfly of the family Danaidae, such as the Monarch or Queen.

diapause A state of suspended animation in which organisms pass through unfavorable environmental conditions.

dimorphism Two different appearances among individuals of the same species; may be seasonal, sexual, or geographic.

discal band A band-shaped marking crossing the discal area of a butterfly wing.

discal cell The large, central cell of a butterfly wing. (See Figure 25.)

dorsolateral Said of markings that are above the lateral (side) markings of (usually) a **caterpillar**.

echinoid Shaped like a sea urchin; said of some insect eggs.

eclose To hatch; to come out of the egg shell; said of **larvae**; also, to come out from a **pupal** case as an adult.

endemic Peculiar to a given region; native, not introduced.

eyespot A ring-shaped or eye-shaped spot; an **ocellus**.

falcate Sickle-shaped; said of forewing tips that are curved and pointed.

family A division of classification; orders are divided into families; families are divided into **genera**.

feces Digestive waste ejected from the **anus**; in insects usually as small pellets.

filament A thread; a slender structure of uniform width.

fulvous Reddish brown; literally, "fox colored."

fusiform Spindle shaped; cylindrical and pointed or rounded at each end.

genitalia The reproductive organs. The external genitalia are much used in advanced classification of butterflies and other insects.

genus, genera A division of classification; **families** are divided into genera; genera are divided into **species**.

glaucous Grayish blue, dusty, like the "bloom" on some kinds of fruit.

gynandromorph An abnormal individual that combines both male and female **characters**.

harpes The paired clasping organs at the tip of the abdomen, as in male **Lepidoptera**; the valves.

hastate Shaped like a spearhead.

heliconian A butterfly of the family Heliconiidae, such as the Gulf Fritillary.

heterocerous Said of **Lepidoptera** in which the **antennae** are of any form other than clubbed.

hilltopping A behavioral trait of such butterflies as congregate on local high points, where males and females meet.

hyaline Transparent or translucent; said of the clear spots found on the wings of insects, such as butterflies.

hybrid A genetic cross between two **species**; usually sterile.

imago The adult insect; also called an imagine.

instar The **larva** at any stage between **molts**. Upon hatching from the egg the larva is in the first instar; after one molt, in the second instar; and so on to the last instar, which transforms into the **pupa**.

labial palpi The paired structures that extend beyond the mouth, between which the **proboscis** (tongue) is coiled.

labium The second pair of jaws on insects. Not noticeable in adult butterflies; forms "lower lip" in **caterpillars**.

lappets The small, flat organs at the base of the front wings, also called **tegulae**, that are sometimes of contrasting color and so used in descriptions.

larva(ae) The second stage of butterfly development; the feeding and growing stage that hatches from the egg; the **caterpillar**. (See Figure 7.)

Law of Priority   A principle of nomenclature that provides that the valid name of an organism is the one that was first applied to it.

Lepidoptera   An order of insects characterized by having minute scales covering the wing surfaces. Includes butterflies, skippers, and moths.

lepidopterist   One who studies members of the order **Lepidoptera**.

lycaenid   A butterfly of the family Lycaenidae, called Gossammer-winged Butterflies.

mandibles   As used here, the chewing jaws of a **caterpillar**.

metamorphic   Said of rocks that have been altered from their original form by heat, water, and great pressure.

metamorphosis   Change in form. Butterflies undergo four stages of metamorphosis: egg, **larva**, **pupa**, and adult (**imago**).

metatibial   Pertaining to the **tibiae** of the hind legs. (See Figure 2.)

middorsal   Applied to markings that are in the central dorsal area, as, for instance, a middorsal line of a **caterpillar**.

migration   A mass movement, usually seasonal, of the individuals of any **species**. Species, the individuals of which migrate, are said to be migratory.

molt   The act of shedding an outgrown body covering; also called ecdysis. This takes place several times in developing **larvae**. The soft new covering permits rapid growth.

monophagous   Feeding on one food; said of caterpillars that feed on only one **species** of plant.

morph   A form. A general, nonspecific term.

Nearctic   Pertaining to the northern regions of North America; often considered as North America north of Mexico.

nectar   The sweet and nutritious material formed by flowers, one of the main foods of adult butterflies and moths. Also a verb, meaning to take nectar from a flower.

nodule   A small lump, node, or swelling.

nuptial flight   A flight in which both males and females take part prior to mating; a mating flight.

nymphalid   A butterfly of the family Nymphalidae.

ocellus   Has two meanings: (1) a single simple eye as distinct from a compound eye; and (2) an eyelike marking on the wing of an insect, such as a butterfly.

ochraceous   Brownish yellow; earth colored.

ochreous   A variant of **ochraceous**.

oligophagous   Feeding on a few species of foods, as, for instance, plants.

osmateria   Odor-producing organs found on the **larvae** of swallow-tail butterflies, protruded from behind the head when the **larvae** are disturbed. Considered protective.

palpi In butterfly adults, the paired structures, either hairy or scaled, found on each side of the **proboscis**. They may be thrust forward or **appressed** to the face.

paradichlorobenzene A chemical, sold as moth crystals, used to prevent infestation of collections by museum pests.

phenol A chemical used in relaxing jars to prevent mold.

pheromone A chemical substance produced by an insect of one sex to attract the other.

phylogenetic Pertaining to the evolutionary relationships of organisms.

pierid A butterfly of the family Pieridae, such as a white or a sulfur.

polymorphic Occurring in more than two forms.

polyphagous Eating many things. Said of **caterpillars** that are able to feed on many **species** of plants ("general feeders").

predator An organism that kills and eats another; for example, some birds and beetles are predators of butterflies. The act of being a predator is called predation.

primaries The anterior pair of wings; the front wings.

proboscis The tubular organ through which adult butterflies sip **nectar**; often called the tongue, but has no anatomical relationship to a mammal's tongue. (See Figure 25.)

prolegs The paired "false legs" along the **abdomen** of a **caterpillar**; these are lost when the caterpillar changes into a **chrysalis**. (See Figure 3.)

pupa(ae) The quiescent, reorganizing stage of a butterfly that follows the **larva**, and from which comes the adult. (See Figure 8.)

pupate To transform into a **pupa**.

radius The radius vein. (See Figure 25.)

Rhopalocera A section of **Lepidoptera** that includes the butterflies and skippers. It refers to the knobbed **antennae**; *rhopalocerous* describes this condition. Once considered a suborder, it no longer is so regarded.

Riker mount A display case made of cardboard, filled with cotton, and covered with glass, that allows handling of insect specimens while being examined. Only one side of the insect can be seen.

riparian An adjective describing plants and animals that live along streamsides. These form a special community.

riodinid A butterfly of the family Riodinidae; a metalmark.

scent pouch An organ containing scent scales (**androconia**), such as is found on the hind wings of a male Monarch.

serpentine A metamorphic rock, usually greenish, with a waxy luster, chemically mostly magnesium silicate; common in parts of California, especially near the coast.

seta(ae) Bristles or spines extending from an insect's skeleton. Soft, slender ones are called hairs or pile; isolated stiff ones may be called spines or **spurs**.

sex-spot A patch of specialized scales, found in some male butterflies. See also **stigma**.

species (sing. and pl.) "A different kind of plant or animal." A population of organisms that reproduces its own kind. Often said to be "a population, the members of which exchange genes."

sphragis A structure attached by the male to the anal end of the female during copulation that prevents further mating; found in butterflies of the genus *Parnassius* (Parnassians).

spinneret The silk-producing organ. Found on the **labium** (lower lip) of the **larva**, it produces silk from which pads, girdles, and **cocoons** are spun.

spiracle A breathing pore through which air enters the body of an insect; on most segments arranged in pairs. (See figures 3, 25.)

spurs Slender appendages attached by a joint to the **tibia** of insects. Depending on the **family**, butterflies may have one or two pairs.

stigma Literally, a mark or brand. In **Lepidoptera**, a patch of scales, usually scent scales (**androconia**), found on the front wings of many skippers and hairstreaks.

subspecies "A taxonomic category less than the species." The only division below a species that has priority standing. Varieties and forms, while useful, have no standing in nomenclature, according to present rules. Subspecies resemble their parent species except in minor points, such as color, and share major characters with other subspecies of their species. A **species** may be divided into two or more subspecies, each occupying a separate range.

sympatric Inhabiting the same geographical area.

tail A projection from the back edge of the hind wing of a butterfly. A tail may be well developed, as in many swallowtails, or very short and slender, as in many hairstreaks.

tarsal claws Curved, paired organs at the outer end of the last tarsal segment of insects, used for clinging to objects. (See Figure 2.)

tawny Lion colored; brownish, sometimes slightly reddish, yellow, darker than **ochraceous**.

taxon (pl. taxa) A taxonomic category into which plants and animals are grouped for purposes of classification.

tegula(ae) The small, flat organs at the base of the front wings; also called **lappets**.

territory The area in which an organism usually lives, and often maintains by defending. Some species of butterflies defend territories; they are said to be territorial.

thecla spot The spot at the lower tip of the hind wing (**tornus**) of a hairstreak butterfly; often contrastingly colored or otherwise conspicuous; absent in some hairstreaks.

thorax The second (middle) section of an insect body. It bears the legs and the wings, if present. (See Figure 25.)

tibia(ae) The fourth section of an insect leg, counting from the base, attached at upper end to the femur and at lower end to the tarsus. It bears the **spurs**, and in some species also small spines. (See Figure 2.)

tibial tuft A tuft of hairs (**setae**) on the **tibiae** of the last (meta-thoracic) leg, used in the classification of certain insects, including Dusky-wings of the genus *Erynnis*.

tornus Hind or anal angle of hind wing; may have tails or special markings.

trachea(ae) The internal hollow tubes into which the **spiracles** open and that carry air to the internal parts of the insect body.

trinomial A scientific name with three parts, which are **genus**, **species**, and **subspecies**. Example: *Incisalia augustus iroides*, the Western Brown Elfin.

tubercle A small bump or mound on the surface of a structure. In **lepidopterous larvae** it may bear hairs or spines. Areas that have tubercles are said to be tuberculate.

urticating Stinging, as nettles, or hairs of some **caterpillars**.

wing cases, wing pads The external appearance of insect wings in a **pupa** (**chrysalis**).

wing cover(s) Same as above, the hard outer skeletal cover(s) of the developing wings in a **pupa** (**chrysalis**).

SOURCES

REFERENCES

The literature on butterflies is very extensive, and only a few titles will be given here. Those interested in further study can get additional titles from the references in some of the general books and papers listed here, or from periodicals, such as the *Journal of the Lepidopterists' Society*, or the *Journal of Research on the Lepidoptera*.

Brower, Lincoln. 1977. "Monarch Migration." *Natural History Magazine*, June–July.

Comstock, John Adams. 1927. *Butterflies of California*. Los Angeles: privately published. Out of print but available in many libraries.

Dornfield, Ernst J. 1980. *The Butterflies of Oregon*. Forest Grove, Oreg.: Timber Press.

dos Passos, Cyril F. 1964. *A Synonymic List of the Nearctic Rhopalocera*. Memoir No. 1. New Haven: The Lepidopterists' Society.

Emmel, Thomas C., and John F. Emmel. 1973. *The Butterflies of Southern California*. Los Angeles: Natural History Museum of Los Angeles County.

Ehrlich, Paul R., and Anne H. Ehrlich. 1961. *How to Know the Butterflies*. Dubuque, Iowa: Wm. C. Brown.

Essig, E. O. 1931. *A History of Entomology*. New York: Macmillan.

Garth, John S. 1950. "Butterflies of Grand Canyon National Park." *Bulletin of the Grand Canyon Natural History Association* 11: 1–52.

Garth, John S., and J. W. Tilden. 1963. "Yosemite Butterflies." *Journal of Research on the Lepidoptera* 2, no. 1: 1–96.

Hartjes, Gloria J. 1980. *Checklist of the Butterflies of Nevada*. Carson City: Nevada State Museum.

Herlan, Peter J. 1962. *A List of the Butterflies of the Carson Range, Nevada*. Carson City: Nevada State Museum.

Hodges, Ronald W., et al., editors. 1983. *Check List of the Lepidoptera of America North of Mexico*. London: E. W. Classey, Ltd., and Wedge Entomological Research Foundation.

Holland, W. J. 1931. *The Butterfly Book*. New York: Doubleday.

Howe, William H. 1975. *The Butterflies of North America*. Garden City, N.Y.: Doubleday.

Klots, Alexander B. 1951. *A Field Guide to the Butterflies*. Boston: Houghton Mifflin.

Miller, Lee D., and F. Martin Brown. 1981. *A Catalogue/Checklist of the Butterflies of America North of Mexico*. Memoir No. 2. The Lepidopterists' Society.

Munz, Philip A. 1959, 1968. *A California Flora and Supplement.* Berkeley and Los Angeles: University of California Press.

Orsak, Larry J. 1977. *The Butterflies of Orange Co., California.* Museum of Systematic Biology Research Series no. 4. University of California, Irvine.

Pyle, Robert Michael. 1974. *Watching Washington Butterflies.* Seattle: Seattle Audubon Society.

———. 1981. *The Audubon Society Field Guide to North American Butterflies.* New York: Knopf.

Scott, James A. 1973. "Mating of Butterflies." *Journal of Research on the Lepidoptera* 11, no. 2: 99–127.

Shapiro, Arthur C. 1975. "The Butterfly Fauna of the Sacramento Valley, California." *Journal of Research on the Lepidoptera* 13, no. 2: 73–82, 115–22, 137–48.

———, Cheryl Ann Palm, and Karen L. Wcislo. 1981. "The Ecology and Biogeography of the Butterflies of the Trinity Alps and Mt. Eddy, Northern California." *Journal of Research on the Lepidoptera* 18, no. 2: 69–151.

Shields, Oakley. 1967. "Hilltopping." *Journal of Research on the Lepidoptera* 6, no. 2: 71–178.

Smith, Arthur C. 1959. *Introduction to the Natural History of the San Francisco Bay Region.* California Natural History Guides, no. 1. Berkeley and Los Angeles: University of California Press.

———. 1961. *Western Butterflies.* Menlo Park, Calif.: Lane Book Co.

Tilden, J. W. 1965. *Butterflies of the San Francisco Bay Region.* California Natural History Guides, no. 12. Berkeley and Los Angeles: University of California Press.

———, and David Huntzinger. 1977. "The Butterflies of Crater Lake National Park." *Journal of Research on the Lepidoptera* 16, no. 3: 176–92.

Tyler, Hamilton A. 1975. *The Swallowtail Butterflies of North America.* Happy Camp, Calif.: Naturegraph.

Urquhart, F. A. 1960. *The Monarch Butterfly.* Toronto: University of Toronto Press.

———. 1976. "Found at Last: The Monarch's Winter Home." *National Geographic* August, 161–73.

Vessel, Matthew F., and Herbert H. Wong, *A Natural History of Vacant Lots.* California Natural History Guides, no. 50. (forthcoming.)

PERIODICALS AND A DIRECTORY

Journal of the Lepidopterists' Society, and *News of the Lepidopterists' Society.* For information, write to the Treasurer of the Society, whose name and address may be found on the inside cover of any current issue.

Journal of Research on the Lepidoptera. For information, write to the

Editor of the Journal. c/o Santa Barbara Museum of Natural History, 2559 Puesta del Sol Road, Santa Barbara, CA 93105.

The Xerces Society, Inc., P.O. Box 3292, University Station, Laramie, WY 82071. Journal: *Atala*; Newsletter: *Wings*. Other publications. Write for information on membership and publications.

The Naturalists' Directory and Almanac (International). 1985. 44th edition. Flora & Fauna Books, 4300 N. W. 23rd Avenue, Suite 100, Gainesville, Florida 32606.

ENTOMOLOGICAL SUPPLY HOUSES

Bio Quip Products. P.O. Box 61, Santa Monica, CA 90406. Carries all types of equipment.

Ianni Butterfly Enterprises (Insect Pins), P.O. Box 81171, Cleveland, OH 44181.

Ward's of California. 11850 Florence Avenue, Santa Fe Springs (L.A.), CA 90670–4490. Carries all types of equipment.

Wind Company, Clo (nets, pins, Riker mounts), 827 Congress Avenue, Pacific Grove, CA 93950.

INDEX OF COMMON AND SCIENTIFIC NAMES

GENERAL INDEX
Including Food Plants

Designer: Rick Chafian
Compositor: G&S Typesetters
Text: 10/12 Times Roman
Display: Helvetica
Printer: Consolidated Printers
Binder: Mt States Bindery

CALIFORNIA NATURAL HISTORY GUIDES

STATEWIDE

Cacti of California, *Dawson*
California Amphibians and Reptiles, *Stebbins*
California Butterflies, *Garth/Tilden*
California Insects, *Powell/Hogue*
California Landscape, *Hill*
California Mammals, *Jameson/Peeters*
Early Uses of California Plants, *Balls*
Edible and Useful Plants of California, *Clarke*
Ferns and Fern Allies of California, *Grillos*
Freshwater Fishes of California, *McGinnis*
Geologic History of Middle California, *Howard*
Geology of Sierra Nevada, *Hill*
Grasses in California, *Crampton*
Introduced Trees of Central California, *Metcalf*
Introduction to California Plant Life, *Ornduff*
Marine Food and Game Fishes of California, *Fitch/Lavenberg*
Marine Mammals of California, *Orr*
Mushrooms of Western North America, *Orr/Orr*
Native Shrubs of Sierra Nevada, *Thomas/Parnell*
Native Trees of Sierra Nevada, *Peterson*
Natural History of Vacant Lots, *Vessel/Wong*
Poisonous Plants of California, *Fuller/McClintock*
Seashore Plants of California, *Dawson/Foster*
Sierra Wildflowers: Mt. Lassen to Kern Canyon, *Niehaus*
Teaching Science in an Outdoor Environment, *Gross/Railton*
Water Birds of California, *Cogswell*

SAN FRANCISCO AND NORTHERN CALIFORNIA

Introduction to the Natural History of the S.F. Bay Region, *Smith*
Introduction to Seashore Life of the S.F. Bay Region and Coast
   of North Calif., *Hedgpeth*
Mammals of the S.F. Bay Region, *Berry*
Native Shrubs of the S.F. Bay Region, *Ferris*
Native Trees of the S.F. Bay Region, *Metcalf*
Rocks and Minerals of the S.F. Bay Region, *Bowen*
Spring Wildflowers of the S.F. Bay Region, *Sharsmith*
Weather of the S.F. Bay Region, *Gilliam*

SOUTHERN CALIFORNIA

Climate of Southern California, *Bailey*
Fossil Vertebrates of Southern California, *Downs*
Introduction to the Natural History of Southern California, *Jaeger/Smith*
Native Shrubs of Southern California, *Raven*
Native Trees of Southern California, *Peterson*
Seashore Life of Southern California, *Hinton*